Olive Oil Cookery

The Mediterranean Diet

by Maher A. Abbas, M.D.
and Marilyn J. Spiegl

Foreword by John W. Farquhar, M.D.
Director, Stanford University Center
for Research in Disease Prevention

Book Publishing Company ◆ Summertown, Tennessee

Cover photo by John Guider
Cover design by Robert Page
Interior design by Barbara McNew

ISBN 0-913990-11-6

01 00 99 98 5 4 3 2

Abbas, Maher A., 1966-
Olive oil cookery : the Mediterranean diet / Maher A. Abbas, Marilyn J. Spiegl : foreword by John W. Farquhar.
p. cm.
"A blend of Mediterranean and California cuisines; more than 150 delicious recipes for a healthier heart".
Includes bibliographical references and index
ISBN 0-913990-11-6
1. Coronary heart disease--Prevention. 2. Cookery (Olive oil) 3. Coronary heart disease--Diet therapy--Recipes. 4. Cookery, Mediterranean. I. Spiegl, Marilyn J., 1930- . II. Title.
RC685.C6A24 1994
641.5'638--dc20 94-17673
CIP

Note: The information in this book is true and complete to the best of our knowledge. All recommendations are made without guarantees on the part of the authors. The authors disclaim all liability in connection with the use of this information.

Calculations for the nutritional analyses in this book are based on the average number of servings listed with the recipes and the average amount of an ingredient if a range is called for. Calculations are rounded up to the nearest gram. If two options for an ingredient are listed, the first one is used. Not included are optional ingredients, serving suggestions, or fat used for frying, unless the amount of fat is specified in the recipe.

TABLE OF CONTENTS

We are indebted to many friends and colleagues who have reviewed our work and tested our recipes. We owe special thanks to the following: Mrs. Dalal Holmin for testing the recipes; to Health Research Design 2000 in Menlo Park, California, for making this work possible; to Nowa Omoigui, M.D., for reviewing this work; to Michel Nabti, Ph.D., for sharing with us his wonderful stories about olive cultivation in the Mediterranean countries, especially Lebanon. Finally, to John W. Farquhar, M.D., for reviewing this work and writing the foreword and Ralph Spiegl, M.D., for his guidance and for providing a description of his methods of growing herbs.

We would like to dedicate our book to three people.

From Maher: to my mother, who is the world's best cook, and to my kind friend Gulla for her help and patience.

From Marilyn: to my greatest friend, my husband Ralph.

FOREWORD

On the culinary journey through life, the prevention of heart attacks can be filled with pleasure. Maher Abbas and Marilyn Spiegl have drawn upon the ancient cultures of the Mediterranean region for practical guidance in the selection of foods that are delicious as well as healthy. Their recipes provide a new *California Cuisine—designed* for health as well as appetizing meals.

The health benefits derive from cooking with olive oil; the authors explain in clear terms how monounsaturated fats (such as olive oil) can help to lower blood cholesterol. That in turn leads to healthier blood vessels, fewer heart problems, and avoidance of stroke. But there is more at stake than prevention of heart disease and stroke. There is increasing evidence that such a cuisine may also prevent certain adult onset cancers (such as breast and colon cancer).

The recipes contained in this book follow all modern guidelines for healthy nutrition, but they do not form a restrictive diet. Gone are the usual feelings of deprivation. I found that just reading through the recipes aroused my appetite, and those we tried in my home produced delicious meals that were very well received by the whole family.

John W. Farquhar, M.D.
Director, Stanford Center for Research in Disease Prevention
Stanford University School of Medicine, California

PREFACE

> *"There is an Arab philosophy about health. They say that health is the digit one, love is zero, glory is zero. Put the one of health beside the others and you are a rich person. But without the one of health, everything is zero."*
>
> *Jack Denton Scott*

During recent years, there has been a revival of interest in olive oil, especially in the United States. Several scientific studies have concluded that this ancient ingredient of fine and ordinary foods, used in the Middle East since prehistoric times, has many important advantages for health. The most important is that it protects against cardiovascular disease, which leads to heart attack and stroke.

I have been interested in olive oil for as long as I can remember. I grew up in Lebanon, one of the countries that has produced and used olive oil down through the ages. Some of my best childhood memories are of the days I spent watching farmers graft young olive trees, harvest the olives, and press them for oil. The traditional method is exciting to watch: man and donkey working together to provide power to the crushing and pressing mills. I also remember sitting under a thousand-year-old olive tree in my grandparents' garden, dreaming the dreams of boyhood, inspired by the beauties of nature so manifest in that tree. But most of all, I vividly recall the dozen jugs placed in our kitchen and filled with the deep green treasure: olive oil. The jugs were filled once a year in the early winter, and I remember how my mother advised us to slow down its consumption in the fall so that we did not run out of whatever oil we had left until the new harvest was produced.

We used olive oil with all meals of the day. For breakfast, it was mixed with a yogurt dish containing herbs (mostly dried thyme), and sometimes with whole olives—the spicy Lebanese olives, so delicious they could be eaten like fruit. For lunch and dinner, it was used in almost every dish. Growing up, I never quite understood why my mother was so obsessed about using it in all of her cooking. Now I think I know why: olive oil is the soul of Mediterranean cuisine, including that of Lebanon. It is a godly gift that was preserved from one generation to another. It is an essential part of daily living and a way to stay in touch with the beauties and bounties of nature. Perhaps these words will mean something to some readers and little to others, but I know that if you grew up in any of those regions that border on the Mediterranean, your heart will skip a beat.

When I immigrated to America, I carried my love for olive oil with me. Having spent most of my early years watching my mother do the cooking, I suddenly found myself the chef in my own kitchen. I began to retrieve from memory the old recipes, but I also tried to create new ones. Later, as an undergraduate at Emory University, I applied my biology and chemistry lessons to an understanding of the human diet and the chemistry of oils. Then I came to Stanford University, where I spent two and a half years in the Falk Cardiovascular Research Center working on cardiovascular diseases, which are the number one killer in America.

A key event in my time at Stanford was meeting Dr. Ralph Spiegl and his wife, Marilyn. They had set up a medical scholarship fund of which I was to become the beneficiary. We hit it off from the beginning. Our relationship soon became more, much more, than that of a generous and hopeful donor and a grateful student. Mutual interests were soon coupled with affection. There is no way to predict or program such things—they just happen. Because of the difference in our ages and levels of achievement, we became much like parents and son. When I first met her, Marilyn—who studied art and education and has held a corporate position as a designer—was already an accomplished cook who had begun to specialize in California cuisine. For her, it was a serious interest. She studied Mediterranean cooking in France and Italy with notable chefs. She taught me a lot, and I reciprocated with a few tips about ordinary, day-to-day cooking as we had done it in Lebanon. Together, we organized several large receptions and dinners, sharing our recipes with friends and families. Before we knew it, we were planning this book.

Maher A. Abbas, M.D.
Stanford University School of Medicine
California

INTRODUCTION

> *"The doctor of the future will give no medicine but will interest his patients in the care of the human frame, in diet, and in the cause and prevention of disease."*
>
> Thomas Edison

Edison may have been too optimistic, but knowledge of nutrition and preventive medicine has grown enormously during the twentieth century. Still, we train our physicians mainly to take care of patients whose troubles stem from a lifetime of bad health habits. This is so unnecessary! Dr. Denis Burkitt wrote recently:

> *"If a floor is flooded as a result of a dripping tap, it is of little use to mop up the floor unless the tap is turned off. The water from the tap represents the cause of disease, the flooded floor the diseases filling our hospital beds. Medical students learn far more about methods of floor mopping than about turning off taps, and doctors who are specialists in mops and brushes can earn infinitely more than do those dedicated to shutting off taps."**

A good example of Dr. Burkitt's point is that cardiovascular diseases, which claim the lives of nearly a million Americans each year, have been attributed mainly to bad health practices

*Burkitt, Denis, "Are our commonest diseases preventable?" *The Pharos*, Winter 1991, pp. 19-21

such as smoking, physical inactivity, and unhealthy eating. There will probably always be a good many people who do not succeed in controlling their appetites and other health habits, but there are encouraging signs too. Consider how many people have quit smoking and the growing popularity of recreational sports and exercise programs. It seems that the more people know about diet, exercise, and all the rest, the more likely they are to improve their behavior.

Another incentive for improving health practices comes from business enterprises and the government. The costs of health care are now about $750 billion per year, which is nearly 15 percent of the gross national product. Either through taxes or directly, it is we, the nation's consumers and common citizens, who ultimately pay for all that. Even though most medical expenses are now funneled through employers and the various levels of government, these costs are born by us through increased costs of goods and services. That gives nearly all of us a personal as well as institutional stake in the health of everyone else. Whatever changes may occur in meeting the expense of national health care, all three entities—private citizens, business, and government—will continue to exert strong pressures for good health practices.

What about doctors and other health care professionals? There the news is mixed. The majority of chronic diseases are known to be caused by unhealthy eating and other mistreatment of our bodies. But doctors' training, and by necessity most of our working hours, are devoted mainly to primary health care, including treatments of various kinds, and high tech diagnostic procedures: treatment rather than prevention. Young and old, rich and poor, patients have been encouraged to think that doctors and pills can fix all of their problems. This leads people to engage in high-risk behavior. Responsible physicians admit that the incentives are very great to do more tests and more surgery than is really good for the patient. We get paid more for that than we will ever get for counseling or educating patients, and we avoid some of the risk of malpractice suits if we order more tests, prescribe more drugs, and even do surgery when it's not entirely certain that it will help the patient.

Some of us who are in the health care business are trying to be responsible leaders of preventive medicine. As we begin to catch up with the desperate need for good primary health care, and as we learn to cope with the new legal climate, I believe we will spend more of our time on education and prevention and less on palliatives or actual cures. Through political and economic pressures, as well as ethical concerns, America will continue to move toward more healthful diets and life-styles. That, I feel certain, is a vision of the future that inspires many newcomers to the profession and will in the end prevail.

Not too many cookbooks begin by lecturing the reader on fine points of diet and health. I don't want to overdo it, but for me it has become a passion to join the two. Medicine is not only a science, but also an art; combining healthful diet and cooking is an art, but with scientifically provable health benefits. In this book, my main purpose is to provide a set of hearty recipes that are of value to any person who cares about delicious and healthful eating. But first, a little information about how to establish and maintain good health, especially cardiovascular health. Then, a little background on the scientific evidence of olive oil's contribution to good health and some information on using it. Following that, the recipes. A special contribution by Ralph Spiegl, M.D., on growing your own herbs appears in the appendix, along with a fascinating look at olive oil's place in history.

THE HEALTHFUL HEART DIET

"The heart moves of itself and does not stop unless forever."

Leonardo da Vinci

This is primarily a cookbook; however, you should remember that your health is influenced by more than your diet. If you are certain that you are living according to good principles of physical and mental health, then perhaps there is no need for you to read this chapter. If, on the other hand, you think a little fine tuning of your basic knowledge about health can be of some benefit, what follows may literally save your life.

Cardiovascular Disease in the United States —

In recent years, cardiovascular and cerebrovascular diseases—those involving blood vessels of the heart and the brain—have become far more common and far more potent. In 1900, such diseases were a factor in less than 13 percent of the total number of deaths in the United States. In 1988, the percentage attributable to cardiovascular diseases was 42 percent.

Not all of this change is attributable to bad health habits. A considerable proportion of total deaths in the earlier period was caused by infectious agents, injuries, and birth abnormalities. Life expectancy is now much greater than it was for our ancestors. More people are now surviving to an age when bad health practices begin to catch up. Consider that 915,000 people—that's 11 large football stadiums filled to capacity—die each year from cardiovascular disease. That includes 510,000 deaths from heart attacks. Another 1.5 million people suffer a heart attack each year but survive.

In total, approximately 66 million people are affected by cardiovascular disease. That's equivalent to the populations of California, New York, Texas, and New Jersey combined. Just think of the consequences: in addition to the deaths, think about the pain and suffering and the many other losses in the quality of life. Consider the economic costs, including the loss of productivity, and how that money could be better spent. Changes in dietary habits have the potential to dramatically improve the health and happiness of our country!

Atherosclerosis is one of the most common cardiovascular diseases in the United States. It results from the hardening of blood vessels, especially arteries, caused by the deposition of cholesterol, fat, calcium, fibrin, and many other substances in their walls. These deposits, also called plaques, occur slowly and progressively over several decades of life, often beginning in childhood. By the age of 40, 80 percent of American men and 65 percent of women have atherosclerotic plaques in their arteries. In contrast to the U.S., these plaques are less common in some other countries—those that generally have a lower standard of living, but a more physically active population, and a more healthful diet.

So plaques are the enemy; exactly what do they do to our bodies? As atherosclerosis progresses, arteries get clogged and lose their flexibility. This leads to a reduction in blood flow to many organs, such as the brain, heart, and kidneys, and of course, raises blood pressure (hypertension). The reduced flow results in organs being deprived of oxygen and essential nutrients. The consequences are many, but the most lethal events are heart attacks and stroke. High blood pressure is a major risk factor in the health of anyone who has the condition. Because of the increased pressure on the walls of blood vessels of the brain, sooner or later the vessels rupture. Although that doesn't always result in instant death, survivors will usually have impaired functions, e.g. slurred speech, paralysis of one whole area of the body, or loss of mental ability.

Unfortunately, hypertension—which affects approximately 60 million Americans—is often not discovered until it is too late to do anything but treat it with medicines. There are no specific signs or symptoms. Many theories have been proposed about causes, including diet, life-style, heredity, and various other medical conditions and diseases. The best evidence about it has to do with diet. Vegetarians, for example, have lower blood pressure than non-vegetarians. And, as pointed out earlier, people in certain areas of the world are much less susceptible.

Coronary Artery Disease and Its Causes —

Your heart is a living pump that beats about 100,000 times a day, 700,000 times a week, 3 million times a month, and about 2.8 billion times in a lifetime! At rest, the pulse rate of a normal person can range between 50 and 80 beats per minute. During exercise, a young athlete can safely sustain a pulse of over 200 beats per minute. At rest, the heart of a trained athlete may beat as slowly as 35 to 50 times per minute. Your heart is the strongest muscle in your body. It pumps about 7,200 liters of blood a day and more than 200 million liters in a lifetime. No human-made pump of its size can begin to match it for efficiency and durability.

Despite what Leonardo had to say, it is now common to stop and restart the human heart in the course of surgery. But when we are born, the vast majority of us have a clean and healthy heart that is ready to undertake the long journey of life without surgical intervention. Unfortunately, some of us forget to take good care of our body and engage in activities that are damaging to our physical as well as emotional well-being. Despite the strength of your heart, you have to keep in mind that it is a delicate organ prone to injury and damage.

Let's consider what happens when coronary arteries become clogged as the result of the process called atherosclerosis. As mentioned earlier, the coronary arteries supply blood to the muscles of the heart. If they are blocked, blood flow is reduced. Depending on the extent of the clogging, reduced blood flow can cause pain, which often resolves after a few minutes (angina), or it can progress to a real heart attack. At the cellular level, portions of heart muscle are deprived of essential nutrients and oxygen. If the deprivation is severe and prolonged, cells begin to die. Depending on how large an area of heart muscle is affected, a person may recover from such an event or die immediately.

A related condition has to do with the heart's electrical system. When blood supply is reduced, the specialized cells which convey electrical signals to the several areas of the heart are often affected because less oxygen is reaching them, causing a condition known as arrhythmia. All of this is discussed in more detail later in this chapter.

Cholesterol —

Why do arteries clog? The risk factors for atherosclerosis are high levels of cholesterol in the blood, hypertension, smoking, diabetes, a family history of heart disease, physical inactivity, stress, obesity, and male gender. Here we are concerned mainly with cholesterol and its effects. There is some confusion about this substance, because the term refers both to an element present in nearly everyone's diet and to a constituent of blood—serum cholesterol.

When it's in the body, cholesterol is a white, soft, fat-like substance. A certain amount of it is necessary to life. It is manufactured in the liver but is also present as a part of many foods, especially from animal sources. Egg yolks, organ meats, any fatty meat, milk and milk products—these are the most common sources, but not the only ones. *Vegetables and their oils have no cholesterol.* However, some non-animal oils such as palm and coconut oils are high in saturated fats and if eaten can increase serum cholesterol.

Serum cholesterol is carried by certain molecules; it cannot dissolve in the blood. These molecules are called lipoproteins. Lipoproteins are, in turn, mostly made of proteins, fats, and cholesterol. There are several types of lipoproteins; the two most important are HDL (high density lipoprotein) and LDL (low density lipoprotein).

Many research studies are being conducted to get a better handle on how HDL and LDL affect atherosclerosis in general, and coronary artery disease in particular. Currently, it is believed that

HDL is mostly produced in the liver and that its role is to circulate in the blood, bind to free cholesterol, and transport it back to the liver where it is eliminated. Simply speaking, HDL's role is to remove cholesterol from the blood and reduce its deposition in the walls of arteries. Therefore, HDL is known as the "good" cholesterol because it slows the formation of atherosclerosis.

HDL "good" cholesterol

Thus, a high level of HDL plays a protective role against atherosclerosis while a low level is believed to have the opposite role. On the other hand, LDL transports cholesterol to the various tissue cells in the body, including the walls of blood vessels such as the coronary arteries. When cholesterol is deposited in the walls of arteries, other substances such as calcium and fats are also deposited, and this leads to its hardening (atherosclerosis). Therefore, LDL is the "bad" cholesterol. A high LDL level is believed to increase the risk of developing atherosclerosis, while a low level is believed to do the opposite.

LDL "bad" cholesterol

HDL levels can increase and LDL levels decrease due to:
- —exercise
- —genetic predisposition
- —a diet high in monounsaturated fats, such as olive oil

Smoking —

By now, everyone knows that smoking is harmful. But many people continue to smoke, even though about 40 percent of deaths from coronary heart disease are linked to smoking, and smokers have more than twice the risk of heart attacks than nonsmokers. In addition to its effects on your heart, other effects of the nicotine and other substances in smoking are:

- —lung cancer
- —emphysema (destruction of the lung)
- —complications in pregnancy; injury to the fetus
- —cancer of the mouth, throat, lung, and bladder

Hypertension —

Hypertension (high blood pressure) is another major risk factor associated with heart disease and stroke. In addition, it can lead to kidney failure and many other health complications. *Know your blood pressure.* Have it checked often if it has ever exceeded the upper limits of the normal range (140/90) and as you get older. There are many options for lowering blood pressure if you are aware that it is getting high, including lower salt intake, better diet, exercise, and weight reduction, as well as medicines.

Blood (or Serum) Cholesterol —

As discussed earlier, serum cholesterol (cholesterol level in the blood) is influenced by two factors: the cholesterol content of your diet and cholesterol synthesis within your liver. Your cholesterol level tends to increase slowly as you age. The table below represents general guidelines of blood cholesterol in adults. These guidelines may change in the future as more research is done, but we are now reasonably certain that an elevated level of serum cholesterol is a major risk factor in the development of atherosclerosis and heart disease.

Average Risk	**Under 200 mg/dl**
Borderline to High Risk	**200-239 mg/dl**
High Risk	**More than 240 mg/dl**
Desired Level	**200 or less mg;/dl**

The above table reflects current recommendations. Readers should know that the above values are arbitrary and that other factors such as LDL and HDL should be taken into consideration when evaluating the risks of an individual. As already discussed, the level of cholesterol in blood is a major risk factor in the development of atherosclerosis and coronary disease. For some people, a genetic pattern may predispose them to a high blood cholesterol level. Get your cholesterol checked at least once every five years. If it measures higher than 200 mg/dl you should test it again to confirm the initial value. If it rises above 240 mg/dl then you should have a complete lipid (fat) analysis. This test should be performed after a fourteen hour fast. A complete analysis measures the levels of LDL, HDL, and triglyceride (another by-product of the body's digestion of fats). An unfavorable level of lipoprotein, or an unfavorable ratio of LDL to HDL can best be attacked through diet and exercise. In some cases, medication is needed.

Diet and Cholesterol —

A good diet begins with an awareness of the content of foods, especially processed foods. Read those labels! Avoid fast foods and avoid eating anything with a high amount of cholesterol, saturated fats, and salt. Here is a list of products to *avoid:*

AVOID:

—whole milk and whole milk dairy products
—butter
—egg yolks
—meat
—pre-mixed salad dressing (unless containing olive oil)
—cream cheese and heavy creams
—mayonnaise and sour cream
—chocolate
—fried foods
—doughnuts and commercial cakes and pastries
—palm and coconut oil

And here is a list of foods you should *eat more* of:

EAT MORE OF:

—fruits and vegetables (but avoid coconut and avocados)
—grains and legumes (beans)
—olive oil in preference to any other form of fat

OLIVE OIL FOR GOOD HEALTH: THE SCIENTIFIC EVIDENCE

> *"An alternative to a low-fat diet for lowering cholesterol levels is the traditional Mediterranean diet, which is just as low in saturated fat and cholesterol. In this diet, olive oil is a major source of energy, fats average 35 to 40 percent of total calories, and rates of coronary disease are as low as in populations with very low-fat diets ... The Mediterranean alternative—using monounsaturated fat as a major dietary component—appears to be at least as healthful, maybe even a better way to improve the lipid profile, and will provide more variety and greater satisfaction to many."*
>
> *Frank W. Sacks and Walter W. Willett*
> *Harvard Medical School, 1991**

In recent years, there has been a great deal of interest in monounsaturated fats and their potential role in reducing the risk of cardiovascular diseases, mainly atherosclerosis. Several studies have been conducted in the U.S. and abroad to investigate the relationship between monounsaturated fats and atherosclerosis.

The basic facts are these. There are three types of fats: saturated (animal sources and coconut and palm oils), polyunsaturated

---

*Sacks and Willett, *The New England Journal of Medicine,* Dec. 12, 1991, Vol. 325, No. 24, pp. 1740-42

(corn, safflower, cottonseed, soybean, and peanut oils), and monounsaturated (olive and canola oil). All three types are found in the human diet. It has been known for several years that a diet rich in saturated fats does increase total blood cholesterol and LDL—the "bad" cholesterol. LDL is believed to promote plaque formation by enhancing cholesterol and lipids (fat) buildup in the walls of arteries. Therefore, a diet high in saturated fat increases blood LDL and cholesterol levels, which in turn increases the risk of developing atherosclerosis and coronary artery disease.

As for monounsaturated and polyunsaturated fats, both are equally inclined to decrease blood LDL, as well as total cholesterol levels. However, polyunsaturated fats lower HDL, a negative health effect. Also, there is evidence associating polyunsaturated fats with cancer. More research is needed, and is being done, to understand these effects.

This much is clear: monounsaturated fats, especially olive oil, have a much better record than other fats in studies of large groups of people. In 1958, the Seven Countries Study was started and continued over 15 years. The goal was to look at coronary heart disease and death from heart attack. The countries studied were the U.S., Finland, Japan, Italy, Yugoslavia, the Netherlands, and Greece. The initial sample involved more than 11,000 middle-aged men. Researchers tracked several variables: diet, cigarette smoking, blood pressure, weight, and exercise habits. The results: Finland and the U.S. had the highest mortality rate from cardiovascular disease; Japan, Greece, and Italy the lowest. When all the information on diet was analyzed, several patterns were apparent: a high intake of saturated fats in Finland and the U.S., a low fat-diet in Japan, but a high-fat diet (mainly rich in monounsaturated fats) in Greece and Italy. Olive oil consumption was the main difference, irregardless of how much was consumed.

Following the Seven Countries Study, some critics questioned the idea that diet was the main reason for the much lower rates of disease in Italy and Greece. Could physical exercise account for the difference? Were the Greeks and Italians more active? Dr. Anna Ferro-Luzzi and her colleagues carried out a study in rural southern Italy to investigate the effect of a change in diet on total cholesterol level, without changing other factors. Animal fats were partially substituted for olive oil for a period of 42 days, without any other changes of life-style. The results were astonishing: in just 42 days, both LDL and total cholesterol blood levels increased significantly.

Recently, several other studies have compared monounsaturated fats, which are synthesized in the body, and other dietary elements and their effects on total cholesterol, LDL, and HDL. In one such study, Drs. Mensink and Katan evaluated the effects of a high-monounsaturated fat and a high-carbohydrate/low-fat diet on total cholesterol and HDL levels. They showed that while both diets equally lowered total cholesterol, HDL fell in the high-carbohydrate/low-fat diet group but did not change in the monounsaturated group. Drs. Mensink and Katan concluded in their report:

> *"The olive oil diet, which combined a high intake of total fat with a low intake of saturated fat, caused a specific fall in non-HDL cholesterol, while leaving HDL cholesterol and triglyceride values unchanged... In view of the supposed anti-atherogenic [prevention of atherosclerosis] effect of HDL, reducing total fat intake per se might not be the best way to prevent CHD [coronary heart disease]."**

In another study, Dr. Grundy compared the effects of a low-fat diet and a high-monounsaturated fats diet with a high-saturated fats diet. As expected, the high-saturated fat diet increases LDL and total cholesterol more than the low-fat diet. In addition, the high-monounsaturated diet did not have any effects on HDL while the low-fat diet decreased it.

What does all this mean? It means that in terms of reducing the risk of developing atherosclerosis, a diet high in fat of the

---

*Mensink and Katan, *The Lancet*, April 25, 1987 pp. 983-984

monounsaturated type, such as olive oil, is probably more healthful than a simple low-fat diet, which is the type of diet recommended by the American Heart Association. *A low fat-diet is good for us in some respects, especially in regard to weight control, but it decreases HDL, and that is believed to increase the risk of atherosclerosis.*

Recently there has been additional interest in olive oil because it contains vitamin E. Many studies have reported that vitamin E may have protective effects against atherosclerosis and cancer because of its anti-oxidizing properties.

Olive oil and cancer: Research studies have also demonstrated that dietary fats contribute to the development of many diet-related cancers, such as breast, colon, prostate, and stomach cancer. The evidence to support this fact has been derived from a number of experiments in the laboratory as well as from population-based studies conducted in the United States and abroad. Some fats are clearly bad for our health and promote cancer. However, can any of them prevent cancer?

In the laboratory, experimental animals fed large amounts of saturated fats from either animal or vegetable sources had a greater chance for developing breast cancer than those that were fed unsaturated fats (such as olive oil). More recently, this laboratory finding was supported by studies conducted with Mediterranean women. In one such study performed by researchers from Spain, Italy, and the Harvard Medical School, women who consumed a high level of olive oil in their diet (rich in monounsaturated fat) were at a significantly smaller risk for developing breast cancer! Furthermore, the higher the amount of olive oil in the diet, the lower the risk of breast cancer.

Olive oil plays a protective role in other types of cancer, such as prostate and stomach cancer. An important study conducted in Israel between 1980 and 1986 showed that non-Jewish Israelis whose diet was high in olive oil had less cancer of the kidneys, bladder, and prostate than Jewish Israelis whose diet was low in olive oil. In another large study in Italy, a country with one of the highest mortality rates from stomach cancer, a high consumption of olive oil in some regions was associated with significantly less stomach cancer.

At present, we don't know exactly why olive oil plays a protective role in preventing cancer, although some researchers believe it may have something to do with the vitamin E and monounsaturated fats abundant in olive oil. As more research is conducted, it will be interesting to see what other health benefits may be associated with olive oil. Certainly more studies are needed, but so far, all investigations have come to the same conclusion: olive oil is indeed a healthful food.

Often we make discoveries only to find out that our ancestors had drawn the right conclusions without the benefits of science. Dr. Ancel Keys once wrote:

> *"I am reminded of Elie Methnikoff, the successor to Louis Pasteur as director of the Institut Pasteur. He became fascinated with longevity and visited Greece on that account. He concluded that centenarians were ten times more common in Greece than in France. We may discount his theory that the credit should go to yogurt, a foodstuff then unknown in other parts of Europe. In any case, from many surveys on the island of Crete, starting in 1957, I have the impression that centenarians are common among farmers, whose breakfast is often only a wineglass of olive oil."**

We should place our trust in science rather than anyone's "impression," but when a doctor's trained observations lead in the same direction as formal research, we must pay attention.

The ultimate value of any scientific discovery or finding is in its application for the good of humanity. It may be too soon to say what the mechanisms are or to answer all the questions about the effects, but when the preponderance of evidence leads to a conclusion that has no known negative effects on health, we would be foolish not to spread the good news and to act on it.

---

*Keys, Dr. Ancel, *The Lancet*, April 25, 1987 pp. 983-984

THE BENEFITS OF A VEGETARIAN DIET

The vegetarian diet has many health benefits that have been recognized for a long time. And as researchers continue to study the population of vegetarians around the world, they are trying to learn more about the protective factors of such a diet against cardiovascular disease and cancer. Clearly a discussion of the subject is far beyond this book, however this short section may be helpful to some readers.

vegans

lacto-vegetarians

lacto-ovo-vegetarians

It is estimated that several million Americans follow a vegetarian diet. Some do so for health-related reasons, while others do for religious, ethical, or philosophical reasons. Since there are a variety of vegetarian styles, let us first define some of the terms that you may hear used by other people when referring to such a diet. *Vegans* eat only plant foods: no product from animal sources is consumed. *Lacto-vegetarians* eat dairy products in addition to plant foods. *Lacto-ovo-vegetarians* eat plant foods, dairy products, and eggs. As you can see, there is a wide diversity among the vegetarian styles. Vegans will have no cholesterol and very little saturated fats in their diet, while lacto-ovo-vegetarians will have some cholesterol and saturated fats. Still however, lacto-ovo-vegetarians in general have less cholesterol and saturated fats in their diet when compared to non-vegetarians.

It is well known that adults who eat vegetarian diets reduce their risks of cardiovascular disease, hypertension, cancer, diabetes, and obesity (overweight). Over the years, many research studies have provided us with evidence supporting the above. The following is a simplified summary of few of these studies. If you are interested in learning more about the subject, please refer to the bibliography section at the end of this book.

As stated before, high blood pressure is one of the major risk factors for development of heart disease and stroke. The higher the blood pressure, the greater is the risk of suffering a stroke. Several surveys conducted earlier this century suggested that meat in the diet may be related to blood pressure. Epidemiologic studies in population groups in the South Pacific Islands, Australia, Malaysia, India, China, Japan, Europe, and the United States showed a trend linking diets containing fish and little meat to low blood pressure. Many questions arose from the above observation: can blood pressure be altered within a population by changing meat consumption? How about blood pressure in vegetarian populations? A group of lacto-vegetarian college students in California were asked to add meat to their diet. Just 11 days later, it was observed that their systolic and diastolic pressures started to increase. In Yugoslavia, a group of elderly persons from the general population were asked to decrease their meat, fish, and egg consumption to five percent of their total caloric intake. After two or more months of changing their diet, a significant decrease in blood pressure was noted. In addition to the above studies, blood pressure in vegetarian populations has been reported to be lower than the blood pressure in the general population. Clearly then, a vegetarian diet does contribute to a decrease in blood pressure which in turn decreases the risk of cardiovascular disease. But the reduction of blood pressure is not the only cardiovascular benefit that a vegetarian diet offers.

Since a vegan diet has no cholesterol and a vegetarian diet contains very little cholesterol and saturated fats, vegetarians have been observed to have lower serum cholesterol. The Framingham Study, which first linked cholesterol and heart disease, reported that the higher the serum cholesterol, the greater the risk of suffering coronary artery disease. Furthermore, the higher the fat intake in a diet, the easier it is to gain weight, even if the total caloric intake is the same. To see how can this is possible, imagine two persons, A and B. Both consume 2,000 calories a day, but person A derives 40 percent of his calories from fat while 10 percent of person B's diet is fat. Assume that both persons have the same weight initially, the same life-style, and the same level of physical activity. Person A will gain more weight than person B. What it boils down to is biochemistry. It takes less energy (calories) to store fat in the body than it does to store carbohydrates (sugars). Therefore, if both persons had an equal excess of calories to store in the body, person A will easily store the fat calories while person B will have to expend some energy to convert the carbohydrates into fat and then store it. Vegetarians as a group tend to be less obese (overweight) than the general population. And we know very well that obese people have an increased risk of high blood pressure and put more work on their heart and kidneys. Furthermore, obese people tend to develop adult onset diabetes, which may lead to heart disease. In that respect, vegetarians enjoy a better "cardiovascular health" when compared to non-vegetarians.

Since non-vegetarians have a greater risk of developing cardiovascular disease than vegetarians, would that risk decrease if a vegetarian diet is adopted? An even more important question is: can someone with already existing cardiovascular disease (i.e. coronary artery disease) decrease the complication of their illness (i.e. heart attacks, chest pain, improvement in physical activity and quality of life) by changing their diet? Obviously, these are complex and difficult questions to answer, especially since cardiovascular disease is a slow, progressing process that may take several decades to manifest. However, there is some evidence that it may be possible to reverse coronary artery disease by changing one's dietary habits and adopting a more "health-sound" life-style.

Recent work by Dr. Dean Ornish has contributed to our understanding of coronary artery disease reversal after life-style changes. In his project "The Life-style Heart Trial," Dr. Ornish reported that it is possible to reverse cardiovascular disease. He took a group of patients with documented heart disease (evaluated by today's medical standards, i.e. angiography, nuclear scans of the heart) and assigned them to two groups: one life-style change group and one control group. The first group was prescribed a low-fat vegetarian diet, moderate physical exercise, stress management training, smoking cessation, and group support. The control group did not make any life-style changes. Both groups were followed for a period of time. A year later, Dr. Ornish observed that the life-style change group had done better than the control group. He reported that he was able to reverse coronary artery disease in some patients in the experimental group. Overall, heart disease continued to progress in the control group. Currently Dr. Ornish is continuing his research and is educating more patients and health care professionals about his health intervention program. So far, at least one insurance company is willing to reimburse fees for patients interested in enrolling in Dr. Ornish's plan, and a few medical centers around the country are considering starting programs modeled after his work. This is very encouraging. How much of the success in reversing heart disease is related to the low-fat vegetarian diet? It is impossible at this moment to assess the relative contribution of diet in this intervention program, however it is clear that a vegetarian diet within the contest of other life-style modifications is indeed beneficial for the cardiovascular well- being of a person and the potential reversal of already-existing disease such as heart disease.

How about cancer? Several epidemiological studies have found in the past that vegetarians have lower rates of certain cancers such as those found in the breast, colon, and prostate. For example, the rate of the mentioned cancers among Seventh-day Adventists is lower than in the general population. Members of this religious group abstain in general from smoking and using alcohol, as well as follow a lacto-ovo-vegetarian diet. But even when cancers linked to smoking and alcohol are accounted for, Seventh-day Adventists still have a lower rate of cancer than the general population. Although their lower cancer mortality may be attributed to no or low meat intake, it is difficult to say how much of it may be due to high intakes of cereal grains, fresh produce such as vegetables and fruits, or avoidance of caffeine. Can the increase in cancer in non-vegetarians be due to the growth factors and hormones fed to animals which are the source of the consumed meat, or can it be linked to some of the preservatives (i.e. nitrites) used to prolong the shelf life of these products? Could the lower rates of cancer among vegetarians be linked to the anti-oxidant effects of vitamins, such as vitamin E, or to the minerals present in fresh produce? As medical researchers try to answer the above questions and determine the relative contribution of different factors to the genesis of or protection against cancer, one fact remains true: the "vegetarian life-style" is healthful and decreases the risk of many life-style related cancers.

In addition to the evidence we have from vegetarian groups such as the Seventh-day Adventists, several studies done on immigrants to the United States from countries such as Japan, Italy, and Greece, have contributed to our knowledge about diet and cancer. For example, it was found that Japanese immigrants have a higher rate of colon cancer when compared to members of their same generation who remained in Japan. It is clear that the life-style changes adopted by the immigrants to the United States are contributing to this increase in cancer incidence and mortality. Some of these changes include an adoption of a more Westernized diet (high in fat and meat consumption), a change in physical activity, and a change in family and social support which may lead to increase of emotional stress. There-fore, it is difficult to know how much of the increase in cancer rate is related to diet and how much is related to the other factors. However, it is clear that diet plays a significant role.

Many factors can lead to the process of disease, and it is often a very challenging task to try to establish simple relationships between cause and effect. However, what we know is that cardio-vascular disease and cancer kill over one million Americans a year. As medical researchers try to gain a better understanding of the causes of heart disease and cancer, it is helpful to know there are ways to protect ourselves and decrease our risk of suffering from life-style related diseases such as the ones mentioned here. The choice of how we live remains ours, and the knowledge of modern nutrition and diet is available to guide us. A vegetarian diet has a lot to offer.

ABOUT OLIVE OIL

Olive Oil Pressings and Grades —

No one has ever taken a census, but I am sure that the majority of olive trees growing today are to be found in close proximity to the Mediterranean Sea. That includes a great many diverse geographical and cultural regions and more than a dozen countries. In the U.S., California and Arizona have also become important growing regions.

The olive tree is an evergreen that can exceed 30 feet in height. It grows in diverse soils, but not in ground that is swampy or damp. Tolerant of a wide range of temperatures, it doesn't thrive in extreme cold, heat, or humidity. It can yield its first olives as soon as three years after planting, but reaches maturity after about thirty years.

Olive trees usually bloom in the spring, with harvest reaching a peak during the fall. Olives are green as they emerge from their buds, turning to purple and ultimately to black. The pulp has about 25 percent oil. The rest is water, acid, and a fibrous residue. Green olives contain less oil, and as they mature and become black, both their oil and acid contents increase. In general, oil from green olives is of a better quality than oil from black olives.

first cold pressing or extra virgin

Once harvested, olives can be cured for eating or pressed for oil. Olives destined for oil are first crushed and then pressed. If pressed with mechanical pressure but without heat, the oil is referred to as the "first cold pressing." Oil derived from a first cold pressing of green olives is referred to as extra virgin oil. By definition (at least for oils imported to the United States from Spain, France, and Italy), the extra virgin grade has less than one percent acidity. Further pressing (second and third) leads to oil with higher acidity and, for most purposes, lower quality. Such oil is adequate for cooking, but will not taste as well in salads, marinades, and other uncooked dishes. Oil that is greater than four percent acidity cannot be used for human consumption.

The grades of oil lower than extra virgin (greater than one percent acidity) are classified as follows:

— 1-1.5 percent - *fine virgin*
— 1.5-3 percent - *regular virgin*
— 3-4 percent - *pure (or virgin)*

olive oil is not regulated by the FDA

There is another distinction in grading olive oil: whether or not it has been filtered. Filtered oil is clear but may have less taste than unfiltered oil, because some of the tastier components have been removed. Labeling can be misleading; since olive oil is not (as of this writing) regulated by the FDA, it's a good idea to know something about the label and the importer. For example, some oils are being sold as extra virgin that meet the established criteria only by adding chemicals during the refining process in order to reduce acidity. "First cold pressing," if honestly used together with "extra virgin," should indicate the highest quality oil.

As a commodity, olive oils are much like wines in several ways: diversity of color, region of growth, and methods of production. There are also subtle differences in flavor and aroma, and (of course!) price. Colors, which vary from yellow to deep green, have nothing to do with quality. There are growing regions that are famous for certain varieties. These orchards ("estates") are known for outstanding quality and even vintage years that are better than others. Like grapes, olives are affected by nuances of soil and climate. When purchasing olive oil, don't assume that the higher the price, the greater your satisfaction; like wines, you should be more concerned with your own subjective response and that of your guests. Don't assume that you require an eight ounce bottle costing $20 for foods that are to be sautéed; save the extra virgin oil for foods that are uncooked (salads, marinades, sandwiches, etc.). Use the regular olive oil to sauté foods.

COOKING WITH OLIVE OIL —

Before you start cooking, we have a few words of advice about the recipes and how to handle olive oil in your kitchen. Your first task is to choose the right oil. Remember what was said earlier: the color of the oil cannot tell you much. Olives change color from green to black as the season goes by. Deep green oil is mostly from the early harvest, and yellow oil is from the end. Any other color comes between the two time periods (but there are exceptions).

The rest is largely a matter of personal taste. For oils that are to be used in uncooked food, it's important to taste several varieties, or at least to take advice. If there is a store near your home that caters to gourmet chefs, ask them to recommend a couple of oils to try; if there is a high quality Greek, Italian, or other Mediterranean kind of restaurant nearby, ask the chef if you may taste different oils. Begin your tasting by smelling. Do this by inhaling the aroma deeply. Then simply take a small amount in a spoon, and place it on your tongue. The back of the tongue is most sensitive to the tastes that discriminate one oil from another. Leave the oil in your mouth for about ten seconds. Then, if you wish to taste another oil, spit out the first sample, clear your palate with a piece of bread, and repeat the test. Just like a wine tasting! If you are left with a bad aftertaste—acidic or fatty—you should try another oil. Remember to look for first cold pressing, extra virgin oil if you wish to use it in uncooked dishes; the less expensive fine or regular virgin oil is perfectly adequate for cooking.

Another consideration is whether the oil has been filtered or not. Filtered olive oil often looks transparent, while unfiltered oil is mostly cloudy. Some people believe that filtering the oil eliminates some of the flavor and deprives the oil of its distinctive quality. Be cautious about labeling; some labels are misleading. For example, "light" olive oil does not mean fewer calories than oil that is not so marked. Rather, this term refers to the flavor of the oil. Light olive oils are produced because Americans in general do not like smelling as a part of tasting and are not used to the distinctive odor of olive oil.

At home, it is best to store olive oil in a glass, glazed clay, or stainless steel container. Don't use iron, copper, or plastic containers. Keep olive oil in a cool and dark place but not in a refrigerator, which would turn it cloudy. Refrigerated oil is also likely to become much more viscous and thick. Before using refrigerated oil, allow it to warm to room temperature.

The recipes that follow are a blend of Mediterranean and California cuisines. They were collected from many sources. Some are authentic to a particular region of the world; others are new and have never before been shared. They have been grouped in order of the courses of a meal. As with any set of recipes, don't hesitate to experiment with variations or to add or subtract ingredients to suit your individual taste.

Olive Oil Cookery

"...of olive oil a hin; and thou shalt make it a holy anointing oil."

Exodus (30:24)

Mangiare non e peccato,
mangiare e una necessita
Saper mangiare, un arte.

Eating is not a sin. Eating is a necessity.
Knowing how to eat is art.

Italian Proverb

Doctors are always working to preserve our health and cooks to destroy it, but the latter are the more often successful.

Denis Diderot (1713-1784)

Appetizers and First Course

Curried Mushroom Caps

Serves 3

Serve hot with a mound of peas in the center of the plate, and border with mushroom caps, *(see* Minted Green Pea Puree, *page 79).*

1½ Tablespoons olive oil

2 teaspoons yellow mustard seeds

1 teaspoon curry powder

1 clove garlic, crushed

1 pound mushroom caps, stems removed and wiped clean

1 teaspoon fresh lemon juice

½ cup white wine

¼ cup vegetable broth

Put the olive oil and mustard seeds in a large frying pan, and heat for 3 minutes on medium heat. Add the curry powder, garlic, mushrooms, lemon juice, white wine, and vegetable broth. Cover and cook over medium heat for 10 minutes, stirring occasionally.

Per serving: Calories: 129, protein: 2 gm., carbohydrates: 7 gm., fat: 7 gm.

Fatee

Serves 6

2 cups non-fat, plain yogurt

2 cloves garlic, minced

½ cup fresh green parsley leaves, finely chopped

2 (15 oz.) cans cooked garbanzo beans

¼ cup olive oil

2 pita breads, cut in 1" pieces

Mix the yogurt with the garlic and parsley; set aside. Empty the garbanzo beans into a medium pot, rinse with fresh water several times, and drain. Add 2 cups of water, and heat for 5 minutes over medium heat. Drain, place on a large plate, and keep warm. Heat the olive oil in a medium frying pan, and toast the pita bread on medium-high heat until crispy and golden, about 5 minutes. Pour the yogurt-parsley mixture over garbanzo beans, and cover with the pita pieces. Serve right away.

Per serving: Calories: 410, protein: 18 gm., carbohydrates: 55 gm., fat: 12 gm.

Simple Scrambled Egg Whites

Serves 2

Scrambled eggs are a typical Mediterranean appetizer.
Serve this with a plate of chopped cucumbers and tomatoes.

½ Tablespoon olive oil

4 egg whites

½ teaspoon dried mint

½ teaspoon allspice

Heat the olive oil in a medium saucepan for 1 minute. Add the egg whites and mint, and scramble over medium heat. Once done, sprinkle allspice over the eggs.

Per serving: Calories: 62, protein: 8 gm., carbohydrates: 1 gm., fat: 3 gm.

Sautéed Mushrooms

Serves 4

A delicious appetizer that can be served for any occasion.

2 Tablespoons olive oil

½ pound fresh mushrooms, chopped

8 cloves garlic, finely minced

⅓ teaspoon black pepper

½ teaspoon ground cumin

salt to taste

¼ cup fresh lemon juice

¼ cup fresh green parsley leaves, minced

Heat the olive oil in a medium pan over medium heat. Add the mushrooms and sauté for 10 minutes. Add the garlic, black pepper, cumin, salt, and lemon juice to mushrooms. Cook 5 more minutes. Remove from the heat and garnish with parsley.

Per serving: Calories: 92, protein: 1 gm., carbohydrates: 7 gm., fat: 7 gm.

Parsley Omelette

Serves 3

6 egg whites

¼ cup fresh green parsley leaves, chopped

3 Tablespoons onion, chopped

¼ teaspoon salt

¼ teaspoon black pepper

⅓ teaspoon cinnamon

2 Tablespoons enriched flour

a dash baking soda

2 Tablespoons olive oil

Place the egg whites, parsley, onion, salt, pepper, cinnamon, flour, and baking soda in a blender, and mix for 4 seconds. Set aside for 10 minutes. Heat the olive oil in a large saucepan over medium heat, and pour in the batter. Lift up the edges of the omelette with a fork, and tilt the pan, allowing the uncooked batter to seep under the omelette. When the omelette is dry on top, fold in half and serve.

Per serving: calories: 133, protein: 8 gm., carbohydrates: 5 gm., fat: 9 gm.

Caponata Alla Siciliana

Serves 4-6

Excellent as appetizer on crackers.

1 (1 lb.) eggplant, peeled and cut in small cubes

1½ Tablespoons olive oil

1 yellow or green pepper, diced

½ large onion, diced

3 tomatoes, finely chopped

1 small stalk celery, diced

1½ Tablespoons capers, well rinsed

3 pitted black olives, sliced

¼ cup vinegar

½ Tablespoon sugar

salt to taste

In a large, deep pan, lightly sauté the eggplants in 1 tablespoon of the olive oil until translucent. Remove from the pan and sauté the peppers in the remaining ½ tablespoon of oil until just softened. Remove from the pan and add to the eggplant. Fry the onion and tomatoes until they are soft. Add the cooked eggplant, peppers, and remaining ingredients. Simmer 10 minutes over low heat, removing from the heat if the vegetables are losing their shape.

Per serving: calories: 111, protein: 1 gm., carbohydrates: 12 gm., fat: 6 gm.

Eggplant with Yogurt

Serves 4

Soaking the eggplant in salty water decreases the amount of oil it absorbs during frying.

1 large eggplant, peeled and sliced ⅓" thick

2½ cups non-fat, plain yogurt

2 cloves garlic, finely minced

¼ cup fresh mint leaves, chopped

2 teaspoons salt

1 cup olive oil

Salt the eggplant slices and soak in water for 15 minutes; place a heavy plate on top of the slices to keep them under the water. Combine the yogurt, garlic, and mint, and set aside. Heat the olive oil in a medium skillet. Remove the eggplant slices from the salty water, and rinse with fresh water several times to remove the salt. Drain. Add the eggplant slices to the olive oil, and fry until golden-brown. Remove the eggplant from the oil; drain for few minutes in a metal strainer or on paper towels. Arrange the eggplant slices on a large plate, and pour the yogurt mixture on top of them. Serve right away.

Per serving: calories: 259, protein: 10 gm., carbohydrates: 22 gm., fat: 12 gm.

Simple Boiled Zucchini with Yogurt

Serves 4

4 medium zucchini

¾ teaspoon salt

1½ cups non-fat, plain yogurt

5 cloves garlic, finely minced

¼ cup olive oil

½ cup fresh parsley, finely chopped

Bring 6 cups of water to boil in a medium pot. Lower heat to medium and add the whole zucchini and ½ teaspoon of the salt. Cook until tender, about 30-45 minutes. Meanwhile, mix the yogurt with the garlic, olive oil, and remaining sea salt. Remove the zucchini from pot, drain well, and slice ¼" to ½" thick. Arrange on a large plate. Cover with the yogurt mixture, and sprinkle with the chopped parsley. Serve right away.

Per serving: calories: 206, protein: 7 gm., carbohydrates: 14 gm., fat: 13 gm.

Steamed Artichokes on French Bread Toast

Serves 6

6 medium artichokes
½ cup freshly squeezed lemon juice
6 cloves garlic, minced
¼ cup olive oil
6 slices French bread, toasted

Strip the leaves off the artichoke hearts. Soak the hearts for about 30 minutes in water with a little lemon juice added to keep the hearts from darkening. Sauté the garlic in 2 tablespoons of the olive oil for 5 minutes. Add the artichoke hearts and the remaining olive oil. Add a little water and steam the artichoke hearts for 20 minutes, or until soft. Serve the hearts on toasted French bread, with the garlic and oil drizzled on it to permeate bread.

Per serving: calories: 212, protein: 5 gm., carbohydrates: 26 gm., fat: 9 gm.

Tomato Omelette

Serves 6

Serve with warm pita bread.

1½ Tablespoons olive oil

1 pound fresh tomatoes, peeled and chopped

2 cloves garlic, finely minced

6 egg whites

3 Tablespoons fresh parsley, finely chopped

salt to taste

¼ teaspoon black pepper, freshly ground

Heat the olive oil in a large saucepan over medium heat. Add the tomatoes and sauté for 4 minutes. Add the garlic and sauté for 3 more minutes. In a bowl, mix the egg whites with the parsley, salt, and pepper. Add the egg mixture to the tomatoes, stir well, and cook uncovered over low heat for 4 minutes. Serve hot.

Per serving: calories: 62, protein: 4 gm., carbohydrates: 4 gm., fat: 3 gm.

Vegetarian Omelette

Serves 3

A traditional omelette from Northern Lebanon.

6 egg whites

3 Tablespoons water

¼ cup fresh parsley, finely chopped

2 Tablespoons fresh basil, finely chopped

1 small red onion, finely chopped

¼ teaspoon salt

¼ teaspoon cinnamon

¼ teaspoon black pepper, freshly ground

2 Tablespoons olive oil

In a small bowl, mix the egg whites with the water, parsley, basil, onion, salt, cinnamon, and pepper. Heat the olive oil in a large skillet over medium heat. Pour the egg mixture into the pan and cook until firm (about 5 minutes), lifting up the edges of the omelette occasionally to let the egg mixture seep underneath and cook.

Per serving: calories: 122, protein: 8 gm., carbohydrates: 3 gm., fat: 9 gm.

Black Olive and Tofu Tarts

Makes 8 tarts

7 cups white onions, thinly sliced

1 Tablespoon olive oil

1 clove garlic, minced

1 package frozen puff pastry, defrosted

1 recipe Tofu Cream Spread (see page 137)

8 black olives, pitted and chopped

Preheat the oven to 400°. Sauté the onions in the olive oil in a large non-stick frying pan until golden-brown, about 25 minutes. Stir often to prevent burning. Add the garlic five minutes before the onions are done; set aside. Roll out the puff pastry ⅛" thick. Cut into 2½" squares and place on a large, greased baking sheet. Spread Tofu Cream Spread on the pastry, and cover with the sautéed onions. Bake in the oven for 15 minutes. Sprinkle the chopped olives on top, and serve.

Per tart: calories: 239, protein: 5 gm., carbohydrates: 22 gm., fat: 14 gm.

Soups

Cauliflower Soup

Serves 5

1 cauliflower head, separated into florets

1 large onion, coarsely chopped

2 small green jalapeño peppers, deseeded and minced (opt.)

2 Tablespoons olive oil

4 cups vegetable broth

salt and pepper to taste

3 Tablespoons fresh lemon juice

2 Tablespoons fresh herbs (thyme, marjoram, etc.)

If you're using jalapeño peppers, use plastic gloves or wash your hands immediately after handling the peppers and seeds. Sauté the vegetables in the olive oil for 5 minutes in a large, deep saucepan. Add the broth, salt, and pepper, and cook until the vegetables are tender (about 12 to 15 minutes). Add the lemon juice and 1 tablespoon of the fresh herbs. Divide up the ingredients into 2 or 3 batches, and puree one batch at a time in a blender; return the batches to the saucepan. Serve garnished with the remaining fresh herbs.

Per serving: calories: 76, protein: 1 gm., carbohydrates: 6 gm., fat: 6 gm.

Cold Yogurt and Mint Soup

Serves 5-6 — Chilling improves the flavor of this soup.

- **2 cloves garlic, minced**
- **1 Tablespoon extra virgin olive oil**
- **salt to taste**
- **4 cups plain, non-fat yogurt**
- **½ cup water**
- **1 large cucumber, peeled and diced**
- **1 cup fresh mint, chopped**

Combine all the ingredients in a medium bowl, and mix for 45 seconds. Chill for 2 hours and serve cold.

Per serving: calories: 145, protein: 13 gm., carbohydrates: 17 gm., fat: 3 gm.

Lentil Soup

Serves 3

This healthful recipe from the mountains of Lebanon is believed to have been around since biblical times. Serve warm or cold.

- **3 cups water**
- **1 cup lentil**
- **½ cup fresh lemon juice**
- **2 Tablespoons extra virgin olive oil**
- **4 cloves garlic, finely minced**
- **½ cup fresh parsley, finely chopped**

In a small pot, bring the water and lentils to a boil. Reduce the heat to medium, cover, and cook for 30 minutes. Drain the lentils and place in a medium bowl. Add the lemon juice, olive oil, and garlic, and combine thoroughly. Top the lentils with the chopped parsley.

Per serving: calories: 326, protein: 16 gm., carbohydrates: 45 gm., fat: 9 gm.

Spinach Soup

Serves 3

1 lb. spinach
3 cups vegetable broth
2 Tablespoons fresh lemon juice
freshly ground black pepper to taste
1 Tablespoon extra virgin olive oil
¼ cup low-fat, plain yogurt
¼ cup Parmesan cheese

Cook the spinach and broth in large saucepan for 15 minutes. Remove from the heat and add the lemon juice, black pepper, and olive oil. Place all in a blender, and puree. Serve in bowls, topped with a dollop of low-fat, plain yogurt and a sprinkling of Parmesan cheese.

Per serving: calories: 128, protein: 7 gm., carbohydrates: 10 gm., fat: 7 gm.

Red Pepper Soup

Serves 7

2 medium onions, coarsely chopped
2 Tablespoons olive oil
12 red peppers, seeded and chopped
5 cups vegetable broth
2 Tablespoons fresh lemon juice
½ teaspoon dried oregano
freshly ground pepper to taste

Slowly sauté the onion in the olive oil over medium heat. Add the chopped peppers and broth. Cover and bring to a low boil, reduce the heat, and simmer until the peppers are cooked (about 20 to 30 minutes). Puree in two batches in a blender, then add the lemon juice and oregano. Chill for 4 hours in the refrigerator to intensify flavors. Add black pepper to taste. Garnish with chopped basil, a spoonful of pesto (page 135) or fresh thyme, if you wish.

Per serving: calories: 77, protein: 1 gm., carbohydrates: 8 gm., fat: 4 gm.

Broccoli Soup

Serves 6

2½ Tablespoons olive oil

2 medium onions, sliced

2 bunches broccoli, cut in 2" chunks (discard tough stems)

4 cups vegetable broth

2 Tablespoons fresh lemon juice

1 teaspoon salt

freshly ground black pepper to taste

2 Tablespoons curry powder

Heat the olive oil in a large skillet over medium-high heat. Sauté the onions for 5 minutes until translucent. Add the broccoli and sauté another 4 minutes to release the flavor. Add the broth and cook covered for 15 minutes until tender. Add the lemon juice, salt, pepper, and curry, and puree in two batches in a blender. Serve hot.

Per serving: calories: 88, protein: 2 gm., carbohydrates: 7 gm., fat: 6 gm.

Carrot Soup

Serves 6

A creamy but very healthy soup.

1 lb. carrots, coarsely chopped

1 large onion, coarsely chopped

1 small unpeeled potato, coarsely chopped

2 Tablespoons olive oil

1 teaspoon fresh sage or thyme

2 cups vegetable broth

2 or 3 Tablespoons fresh ginger, peeled and chopped

2 Tablespoons peanut butter

2 Tablespoons fresh lemon juice

salt and pepper to taste

Sauté the vegetables in the olive oil over medium heat for 5 minutes to release their flavors. Add the herbs and broth, and cook until soft, about 15 to 20 minutes. Puree in two batches in a blender. Add the ginger, peanut butter, lemon juice, salt, and pepper. Mix well and serve hot.

Per serving: calories: 135, protein: 2 gm., carbohydrates: 15 gm., fat: 6 gm.

Lentil and Bulgur Soup

Serves 8

2 cups uncooked lentils

¼ cup olive oil

2 large white onions, finely chopped

½ cup bulgur (cracked wheat)

¼ cup lemon juice

¼ teaspoon cumin

¼ teaspoon cinnamon

¼ teaspoon black pepper

soy sauce to taste

Rinse the lentils, place in a large pot, and add 9 cups of water. Bring to a boil, reduce heat to medium and cook covered for 15 minutes. Heat the olive oil in a medium saucepan, and sauté onions for 10 minutes on high heat. Add the onions to the lentils along with the bulgur, lemon juice, cumin, cinnamon, black pepper, and soy sauce, cover, and cook for 30 minutes. Serve hot.

Per serving: calories: 216, protein: 9 gm., carbohydrates: 30 gm., fat: 7 gm.

Summer Lentil and Lemon Soup

Serves 5

1 cup uncooked lentils

3 Tablespoons olive oil

½ large white onion, finely diced

4 cloves garlic, finely minced

½ cup fresh lemon juice

½ teaspoon ground cumin

½ cup radishes, sliced

¼ cup fresh coriander, chopped

Bring 4 cups of water to boil in a medium pot. Add the lentils, reduce the heat to medium, cover, and cook for 30 minutes. Drain. In a small saucepan, heat 2 Tablespoons of the olive oil, and sauté the onion over high heat until golden. Add the onions, garlic, lemon juice, remaining olive oil, and cumin to lentils, and mix well. Serve in bowls, topped with a sprinkling of radish slices and chopped fresh coriander.

Per serving: calories: 194, protein: 8 gm., carbohydrates: 22 gm., fat: 8 gm.

Tuscan Bean Soup

Serves 8-10

This is best made the day before so that the flavors can marry. Make it on a lazy weekend day.

2 cups dried cannellini or white beans, soaked overnight

¼ cup olive oil

2 carrots, finely chopped

2 celery sticks, finely chopped

2 leeks, finely chopped

3-4 ripe tomatoes, seeded and mashed

3 cloves garlic, crushed

2 sprigs fresh thyme

salt and black pepper to taste

6 slices stale bread

5 cups vegetable broth

1 cup Savoy cabbage, shredded and steamed

1 cup red onions, sliced (opt.)

Boil the beans in a large amount of water until slightly softened. Cover, set aside for 1 hour, and drain. Puree ¾ of beans in an equal amount of fresh water, and reserve the rest. Sauté the carrots, celery, and leeks in the olive oil in a large saucepan. When they are soft, add the tomatoes, garlic, thyme, salt, and pepper. Cook 10 minutes, then add the bean puree. Cook for 1 hour, adding more broth if the soup seems too thick. About 10 minutes before the end of cooking, stir in the whole beans to heat thoroughly. Ladle the soup over the dry, sliced bread in individual bowls, and top with the cooked cabbage on top. Sliced red onions make a good accompaniment.

Per serving: calories: 264, protein: 11 gm., carbohydrates: 41 gm., fat: 5 gm.

Vegetable Soup

Serves 8

3 Tablespoons olive oil

1 cup red onion, finely chopped

½ green pepper, chopped

1 carrot, sliced

2 potatoes, peeled and finely chopped

1 cup fresh spinach, chopped

2 large tomatoes, peeled and chopped

3 teaspoons fresh parsley, chopped

¼ teaspoon black pepper

¼ teaspoon paprika

3½ cups water

In a medium pot, sauté the onion, green pepper, and carrot in the olive oil over medium heat for 5 minutes. Add the potatoes, spinach, tomatoes, parsley, black pepper, paprika, and water to the pot, cover, and cook over medium heat for 40 minutes. Serve hot.

Per serving: calories: 92, protein: 1 gm., carbohydrates: 11 gm., fat: 5 gm.

Creamy Soup

Serves 3

½ Tablespoon olive oil

1 medium white onion, diced

2 cloves garlic, minced

1 cup vegetable stock

¼ teaspoon Tabasco sauce

4 tomatoes, diced

1½ cups tofu, crumbled

3 Tablespoons fresh parsley leaves, minced

Heat the olive oil in large skillet over medium heat. Add the onion and garlic, and sauté 5 minutes. Add the stock, Tabasco sauce, and tomatoes, and cook for 5 more minutes, stirring constantly. Pour into a blender, add the crumbled tofu, and blend until smooth. Serve cold with a sprinkling of minced parsley.

Per serving: calories: 164, protein: 10 gm., carbohydrates: 13 gm., fat: 8 gm.

Tuscan Bread Soup with Tofu

Serves 8

1 small red onion

2 carrots

2 stalks celery

3 Tablespoons olive oil

2 (28 oz.) cans pureed tomatoes (about 7 cups)

8 cups dried Italian and French bread, cubed

1 cup soft tofu, drained, dried, and cubed

2 cloves garlic, minced

2 Tablespoons dried basil

Parmesan cheese to taste (opt.)

Chop the vegetables into ¼" pieces. Heat 2 tablespoons of the olive oil in a large soup pot over medium heat. Add the vegetables and saute until crisp. Add the pureed tomatoes and bring to a boil, then add the bread and tofu, and simmer for 15 minutes. Add the rest of the olive oil, the garlic, and the basil one minute before the soup is done. Ladle into soup bowls and sprinkle with Parmesan cheese, if desired. Serve right away.

Per serving: calories: 239, protein: 7 gm., carbohydrates: 34 gm., fat: 7 gm.

Tofu and Vegetable Soup

Serves 5

2 Tablespoons fresh ginger, grated

3 Tablespoons soy sauce

6 cups water

½ cup green onions, chopped

2 small carrots, sliced

1 small white onion, finely chopped

6 cloves garlic, minced

3 Tablespoons olive oil

2 cups mushrooms, sliced

1 cup cauliflower, chopped

½ pound medium-firm tofu, drained and cubed

1 cup green beans, chopped

1 small zucchini, sliced

2 cups spinach, chopped

¼ cup white miso

In a large pot, simmer the ginger, soy sauce, and water for 20 minutes. Strain and reserve the liquid. In a large skillet, sauté the green onions, carrots, white onion, and garlic in the olive oil over medium heat for about 10 minutes. Add the reserved liquid and bring to a boil. Add the mushrooms, cauliflower, tofu, beans, and zucchini, reduce the heat and simmer for 5 minutes. Decrease the heat to low, add the spinach and miso, and cook for 1 minute. Serve right away.

Per serving: calories: 192, protein: 8 gm., carbohydrates: 16 gm., fat: 9 gm.

Eggplant Tofu Soup

Serves 8

1 cup white onions, minced

1½ cups celery, minced

1½ cups potatoes, diced into ¼" cubes

1 large eggplant, peeled and diced into ¼" cubes

¼ cup olive oil

1 teaspoon curry powder

1 teaspoon dried oregano

2 Tablespoons fresh basil

4 cups tomato juice

1 cup soft tofu, drained, dried, and cut into 1" cubes

In a large soup pot, sauté the onions, celery, potatoes, and eggplant in the olive oil over low heat until tender, about 25 minutes. Add the curry powder, oregano, basil, tomato juice, and tofu, and cook an additional 15 minutes. Puree in a blender in several batches, and return to the pot. Pour into warm bowls and serve right away.

Per serving: calories: 159, protein: 4 gm., carbohydrates: 18 gm., fat: 7 gm.

Pasta, Beans, Rice, and Polenta

Hummus

Serves 5

Traditionally, this is served on one large platter into which everyone dips their wedges of pita bread. Accompany with olives and cucumber pickles.

2 (15 oz.) cans cooked garbanzo beans, drained and washed

⅓ cup tahini

½ teaspoon salt

3 cloves garlic, minced

1 teaspoon extra virgin olive oil

½ cup fresh lemon juice

¼ cup fresh parsley, chopped

Place the garbanzo beans in a food processor, and blend until smooth, adding water if the beans seem difficult to process. Add the tahini and process until well mixed. Place the mixture in a large bowl. Add the salt, garlic, olive oil, and lemon juice to the garbanzo beans, and mix for 1 minute. Spread a large plate, and sprinkle with the chopped parsley.

Per serving: calories: 395, protein: 16 gm., carbohydrates: 55 gm., fat: 11 gm.

Fettucini with Yogurt Sauce

Serves 4

1 lb. fettucini

6 cloves garlic, minced

2 Tablespoons extra virgin olive oil

2 Tablespoons fresh parsley, finely chopped

1 cup non-fat, plain yogurt

Cook the fettucini according to the package instructions, then drain well. Mix the garlic with the olive oil, parsley, and yogurt. Pour the yogurt mixture on the fettucini, and mix well. Serve right away.

Per serving: calories: 265, protein: 10 gm., carbohydrates: 40 gm., fat: 7 gm.

Spaghetti with Parmesan Cheese

Serves 6

1 lb. spaghetti

5 cloves garlic, minced

¼ cup Italian seasoning (a combination of fresh basil, oregano, thyme, and parsley)

3 Tablespoons extra virgin olive oil

½ cup Parmesan cheese

Prepare the spaghetti according to the package instructions. Drain and place in a large bowl. Combine the garlic, Italian seasoning, and olive oil. Add to the spaghetti, and mix well for 1 minute. Sprinkle with Parmesan cheese, and serve piping hot.

Per serving: calories: 202, protein: 7 gm., carbohydrates: 24 gm., fat: 7 gm.

Maher's Favorite Garbanzo Beans

Serves 4

2 cloves garlic, finely minced

salt to taste

½ cup fresh parsley, minced

2 Tablespoons extra virgin olive oil

1 Tablespoon water

½ cup fresh lemon juice

2 (15 oz.) cans cooked garbanzo beans

Combine the garlic, salt, parsley, olive oil, water, and lemon juice in a small bowl. Drain the garbanzo beans, rinse several times with fresh water, and place in a medium bowl. Add the parsley mixture to the garbanzo beans and mix well.

Per serving: calories: 416, protein: 16 gm., carbohydrates: 62 gm., fat: 9 gm.

Pasta with Walnut Sauce

Serves 6

1 lb. fettucini

3 cloves garlic

2 Tablespoons fresh whole wheat bread crumbs

1 cup low-fat milk

salt and pepper to taste

3 Tablespoons extra virgin olive oil

1 cup whole walnuts

6 leaves fresh basil

Cook the fettucini according to the package instructions, and drain. Process the garlic, bread crumbs, milk, salt and pepper, and olive oil in a food processor or blender until the consistency of cream. Pour the sauce over the hot pasta, and toss. Garnish with the whole walnuts and basil leaves, and top with Parmesan cheese, if desired. You can reduce the fat content by one-third by cutting the amount of walnuts to ½ cup.

Per serving: calories: 318, protein: 8 gm., carbohydrates: 29 gm., fat: 18 gm.

Risotto

Serves 6

Italian arborio rice is simply the best type of rice for a creamy, yet firm, risotto. You could substitute with California pearl rice.

1½ cups Italian arborio rice
2 Tablespoons olive oil
1 red onion, finely chopped
½ cup mushrooms, chopped
1 bay leaf
1 cup sweet white wine
5 cups vegetable broth
1 cup Parmesan cheese
½ cup fresh parsley, chopped

In a large pan, sauté the rice in olive oil for 2 minutes, coating the grains completely. Add the red onion, mushrooms, bay leaf, and wine. As the rice cooks, add 4 cups of the broth. Add a little more broth or water if the broth is absorbed before the rice is cooked (about 25 minutes). When the rice is al dente, stir in the Parmesan cheese. Sprinkle with the chopped parsley, and serve.

Per serving: calories: 245, protein: 9 gm., carbohydrates: 28 gm., fat: 9 gm.

Riz Bi-Sheeree

Serves 6

A tasty Lebanese rice recipe that can be served with any vegetable recipe and a green salad. Or you can simply serve it with a fresh, low-fat yogurt.

¼ cup olive oil

1 cup wheat semolina, or 1 cup uncooked angel hair pasta, broken in quarters

2 cups uncooked rice

½ teaspoon salt

Heat the olive oil in a medium pot, and sauté the semolina or pasta over medium heat until brown. Add the rice, mix well, and sauté for 4 minutes. Add the salt and 5 cups of water. Increase the heat to high, and bring to a boil; then reduce the heat to medium, cover, and simmer for about 20 minutes.

Per serving: calories: 305, protein: 6 gm., carbohydrates: 48 gm., fat: 9 gm.

Tomato and Rice Stew

Serves 5

2 Tablespoons olive oil

4 small red onions, finely chopped

5 cups tomatoes, chopped

½ teaspoon white pepper

1 teaspoon cinnamon

¼ teaspoon black pepper

1 cup uncooked rice

¼ teaspoon sea salt

Heat the olive oil in a large pot, and sauté the onion over medium heat for 10 minutes. In a food processor (or several batches in a blender), puree the tomatoes for 1 minute. Add the tomatoes, white pepper, black pepper, and cinnamon to the onions, and bring to a boil. Add the rice and combine. Cover, reduce the heat to low, and simmer for 30 minutes. Serve hot.

Per serving: calories: 202, protein: 4 gm., carbohydrates: 33 gm., fat: 6 gm.

Spaghetti with Fresh Tomatoes

Serves 4

An excellent Italian cook in Sienna told Marilyn about a great tip for pasta: Put the cooked, drained pasta in the sauce you will be using for 2 minutes before serving to heat it thoroughly, and serve piping hot on warm plates. She also told Marilyn that Italians do not speak for 15 minutes while they are eating their hot pasta!

1 lb. spaghetti

2 Tablespoons olive oil

4 cloves garlic, minced

3-4 cups tomatoes, chopped (4 large)

1 teaspoon thyme

1 teaspoon fresh parsley, finely chopped

½ teaspoon black pepper

Cook the spaghetti according to the package instructions, then drain. Heat the olive oil in a pan, and sauté the garlic for 5 minutes over medium heat. Add the tomatoes, thyme, parsley, and black pepper to the garlic, and mix. Cover the pan and simmer for 10 minutes. Pour the tomato mixture over the spaghetti, and combine well. Serve hot.

Per serving: calories: 259, protein: 7 gm., carbohydrates: 41 gm., fat: 7 gm.

Baked Polenta with Goat Cheese

Serves 4

¾ cup yellow cornmeal
½ teaspoon sea salt
¼ cup goat cheese, softened
2 Tablespoons olive oil
¼ teaspoon black pepper, freshly ground
a sprig of fresh rosemary
several sprigs of rosemary for garnish

Preheat the oven to 400°. Combine the cornmeal, salt, and 2 cups of water in a 2-quart casserole, and bake for 20 minutes. Chop the sprig of rosemary, and rub it between your hands to warm it and release its aromatic oils. Add it to the cornmeal along with the goat cheese, olive oil, and pepper. Bake an additional 25 minutes. Remove from the oven, and let stand for 3 minutes. Garnish with the remaining springs of rosemary, and serve hot.

Per serving: calories: 191, protein: 4 gm., carbohydrates: 21 gm., fat: 10 gm.

Curried Pilaf

Serves 4

Indian basmati rice is flavorful and perfect for pilafs.

1 cup basmati rice
1 Tablespoon olive oil
3 cloves garlic, sliced
1 Tablespoon curry powder
½ cup currants or raisins
1 cup onion, finely chopped
1½ cups vegetable broth
1 teaspoon salt
¼ teaspoon freshly ground black pepper
½ cup slivered almonds

Wash and drain the rice. Sauté the olive oil and sliced garlic in a medium pot for 2 minutes. Add the washed rice, curry powder, currants or raisins, and chopped onion. Stir to coat the rice over low heat for 4 minutes. Add the broth and bring to a boil. Cover and cook for 15 minutes. Add the salt and pepper, and let stand for 5 minutes. Garnish with the slivered almonds, and serve. You can reduce the amount of fat almost in half by cutting the amount of almonds to ¼ cup.

Per serving: calories: 335, protein: 7 gm., carbohydrates: 48 gm., fat: 12 gm.

Rice and Lentil Mudjera

Serves 6

Serve with Tomato Salad *(see page 105).*

1 cup uncooked lentils
1 cup uncooked rice
¼ teaspoon salt
¼ teaspoon cinnamon
2 medium white onions, chopped
2 Tablespoons olive oil
½ cup fresh lemon juice

Soak the lentils and rice in separate pots in enough water to cover for 20 minutes. Drain the lentils and place them in a medium pot. Add enough water to cover ½ inch above the lentils, and simmer over medium heat until tender, about 20 minutes. Keep the level of the water in the pot above the lentils as they cook. Drain the soaked rice and add it to the lentils, along with the salt and cinnamon and 2 more cups of water. Bring this to a boil, then reduce the heat, cover, and cook for 20 minutes. While the lentils and rice are cooking, sauté the onions in the olive oil until golden-brown. When the lentils and rice are done, spread the onions on top, and sprinkle with the lemon juice. Serve right away.

Per serving: calories: 236, protein: 8 gm., carbohydrates: 39 gm., fat: 5 gm.

Spinach Pilaf

Serves 4

- **2 Tablespoons olive oil**
- **½ cup onion, finely chopped**
- **1½ cups basmati rice**
- **3 cups vegetable broth**
- **¼ cup slivered almonds**
- **1 teaspoon olive oil**
- **⅔ cup spinach washed, stemmed, and finely chopped**
- **⅓ cup fresh parsley, chopped**
- **¾ teaspoon salt**
- **¼ teaspoon freshly ground black pepper**

Heat the olive oil in a large, heavy saucepan over medium-high heat. Add the chopped onion and sauté until translucent, about 5 minutes. Add the rice and mix with the oil and onion until all the grains are coated, about 2 minutes. Add the broth and bring to a boil. Stir, turn the heat to very low, and cover tightly. Simmer for about 15 minutes without stirring or lifting the cover. While the rice is cooking, sauté the slivered almonds in the teaspoon of olive oil for 2 minutes. Pre-heat the oven to 400°. Add the spinach, parsley, salt, pepper, and sautéed almonds to the cooked rice. Place in an attractive, oven-proof serving dish, and heat in the oven for 5 minutes; serve.

Per serving: calories: 337, protein: 6 gm., carbohydrates: 48 gm., fat: 13 gm.

Lebanese Spinach with Rice and Pine Nuts

Serves 6

2 cups uncooked rice

3 Tablespoons olive oil

½ cup pine nuts

½ large white onion, finely chopped

3 cups fresh spinach, finely chopped

1 teaspoon Dijon mustard

¼ teaspoon black pepper

½ teaspoon salt

½ cup fresh lemon juice

Simmer the rice in 4 cups of water until the water has been absorbed and the rice is tender. In a medium pan, heat 1 tablespoon of the olive oil, and sauté the pine nuts over medium heat until golden. Set aside. In a large skillet, heat the remaining 2 tablespoons of olive oil, and sauté the chopped onion for 4 minutes. Add the spinach, mustard, pepper, and salt, and sauté over medium heat for 12 minutes. In a large bowl, mix the rice, pine nuts, and the spinach mixture. Sprinkle with the fresh lemon juice, and serve hot.

Per serving: calories: 339, protein: 8 gm., carbohydrates: 47 gm., fat: 13 gm.

Spring Risotto

Serves 6

- **1 onion or leek, finely chopped**
- **2 Tablespoons olive oil**
- **2 cups zucchini, diced**
- **1 cup asparagus or green pepper, chopped**
- **½ cup carrots, diced**
- **1 cup tomatoes, diced**
- **salt and pepper to taste**
- **1½ cups uncooked Italian arborio rice**
- **4 cups hot vegetable broth**
- **1 cup white wine**
- **3 Tablespoons extra virgin olive oil**

In a large saucepan, sauté the onion or leek in 2 tablespoons of the olive oil until it begins to brown. Add all the other vegetables, except the tomatoes, and cook for 10 minutes. Add the tomatoes, salt, and pepper, and cook 15 minutes more. Add the rice, a little of the hot broth, and the white wine. Continue adding ½ cup of the hot broth every few minutes until the rice absorbs the liquid. It takes approximately 20 to 30 minutes for the rice to cook to the al dente stage. Stir in 3 tablespoons of olive oil to flavor, and serve with Parmesan cheese.

Per serving: calories: 292, protein: 4 gm., carbohydrates: 36 gm., fat: 11 gm.

Rice Stew

Makes 10 cups

1 Tablespoon olive oil

1 large white onion, chopped

2 (8 oz.) cans tomato sauce

6 cups water

½ teaspoon cumin

¾ teaspoon black pepper

¼ teaspoon cinnamon

½ teaspoon salt

½ cup rice

½ cup fresh parsley, chopped

Heat the olive oil in a large pot, and sauté the onions over medium heat for 7 minutes. Add the tomato sauce, water, cumin, black pepper, cinnamon, and salt to the onions, and mix. Bring to a boil, stirring occasionally. Add the rice, reduce heat to low, cover, and simmer for 25 minutes. Add the parsley about 5 minutes before the rice is done. Serve hot.

Per cup: calories: 62, protein: 1 gm., carbohydrates: 11 gm., fat: 1 gm.

Spicy Tofu

Serves 6

This dish is the perfect topping for a bed of rice and can be added to stir-fried vegetables, or you can try it cold with a dash of fresh squeezed lemon juice. Nutritional yeast is not to be confused with brewer's yeast, which is not nearly as tasty. Nutritional yeast is golden in color and has a mild, cheese-like flavor.

2 lbs. tofu

3 Tablespoons olive oil

½ teaspoon turmeric

1 teaspoon dill weed

½ teaspoon basil

½ teaspoon powdered thyme

½ teaspoon salt

½ teaspoon curry powder

½ teaspoon ground cumin

1 glove garlic, pressed

2 Tablespoons soy sauce

¼ cup nutritional yeast

Slice the tofu ½" thick, firmly pat out excess water with a paper towel, and dice the slices into small cubes. Heat the olive oil in a large skillet on high heat. Add the diced tofu and sauté until lightly browned, about 5 minutes. Drain any extra liquid that is released as the tofu cooks. Reduce the heat to medium, add the turmeric, and stir until the tofu is evenly yellow. Add the dill weed, basil, thyme, salt, curry, and cumin, stirring well after each addition. Add the garlic and then the soy sauce, stirring constantly. Add the nutritional yeast just before serving, and mix well.

Per serving: calories: 195, protein: 13 gm., carbohydrates: 5 gm., fat: 12 gm.

Mushroom Chili

Serves 6

3 Tablespoons olive oil

4 cups assorted mushrooms (portobello, oyster, shiitake), chopped

2 cups onions, diced

1½ cups celery, diced

1½ cups Italian plum tomatoes, cut up

1 cup tomato soup

1 cup textured vegetable protein granules

2 cups cooked red kidney beans

2 Tablespoons chili powder, or more to taste

1 Tablespoon dried oregano

Heat the olive oil in a large skillet over medium heat. Add the mushrooms, onions, and celery, and sauté until tender. Add the tomatoes, soup, textured vegetable protein, beans, chili powder, and oregano, and cook over low heat for 45 minutes. Serve with corn bread.

Per serving: calories: 232, protein: 13 gm., carbohydrates: 28 gm., fat: 7 gm.

White Bean Ragout

Serves 4

1 cup dry white beans

4 teaspoons salt

2 Tablespoons olive oil

1 medium yellow onion, chopped

4 cloves garlic, minced

2 cups tomato juice

1 cup textured vegetable protein granules

¼ teaspoon ground red pepper

salt and pepper to taste

In a medium pot, combine the beans with 5 cups of water and 2 teaspoons of the salt. Heat to boiling and cook for 2 minutes. Turn off the heat, cover, and let stand for one hour. Drain any excess water. Add 5 cups of water and the rest of salt, and simmer for 45 to 60 minutes until tender. Heat the olive oil in a medium soup pot. Add the onion and sauté for 10 minutes. Add the garlic and cook for 1 more minute. Add the cooked beans, tomato juice, textured vegetable protein, and red pepper, cover, and simmer for 30 minutes. Add salt and pepper to taste. Serve with rice on the side.

Per serving: calories: 291, protein: 18 gm., carbohydrates: 38 gm., fat: 7 gm.

Spinach Garbanzo Beans

Serves 4

2 lbs. fresh spinach, chopped

2 Tablespoons olive oil

2 cups yellow onions, chopped

2 cloves garlic, minced

½ cup textured vegetable protein granules

½ cup cilantro, chopped

2 teaspoons cumin

1½ cups cooked garbanzo beans

½ cup lemon juice

salt and pepper to taste

Wash the spinach and drain slightly. Heat the olive oil in a large pot over medium heat. Add the onions and cook for 5 minutes. Add the garlic, textured vegetable protein, and cilantro, and cook for an additional 5 minutes. Add the drained spinach, mix well, and cook partially covered until soft, about 5 minutes. Stir in the cumin and garbanzo beans, and simmer for 5 minutes. Stir in the lemon juice 2 minutes before the end of the cooking time. Add salt and pepper to taste. Serve with rice on the side.

Per serving: calories: 273, protein: 15 gm., carbohydrates: 34 gm., fat: 7 gm.

Wild Rice with Dried Fruits

Serves 5

1 cup celery, finely chopped

1 cup yellow onions, finely chopped

2 Tablespoons olive oil

1⅔ cups wild rice

¼ cup textured vegetable protein granules

⅔ cup white wine

1½ cups water

2 teaspoons fresh parsley, finely chopped

½ cup dried apricot, chopped in ¼" pieces

½ cup dried prunes, chopped in ¼" pieces

¼ cup walnuts, finely chopped

¼ cup almonds

1 teaspoon fresh sage, or ½ teaspoon crumbled dried sage

Preheat the oven to 350°. Sauté the celery and onions in the olive oil in a medium sauté pan until crisp. Mix the rice with the textured vegetable protein. In a large baking dish, combine the rice mixture with the wine, water, cooked celery, onions, and remaining ingredients. Mix well, cover, and bake for 1 hour, stirring occasionally. After one hour, check to see if the rice is tender; if not, add ½ cup water and bake an additional 15 minutes.

Per serving: calories: 408, protein: 10 gm., carbohydrates: 58 gm., fat: 11 gm.

Vegetables

Cauliflower with Tahini

Serves 6

2 Tablespoons olive oil

2 pounds cauliflower, separated into small flowerets

¼ cup tahini

2 cloves garlic, minced

2 Tablespoons water

1 cup fresh lemon juice

Heat the olive oil in large skillet over medium heat. Add the cauliflower and sauté for 15 minutes. Mix the tahini, garlic, water, and lemon juice in a small bowl. Add this mixture to the cauliflower, stir for 1 minute, and cook for 5 more minutes, uncovered.

Per serving: calories: 151, protein: 3 gm., carbohydrates: 12 gm, fat: 9 gm.

Green Beans Alla Italiana

Serves 3

¾ pound thin green beans, cut in 2" pieces

1 teaspoon garlic, minced

1½ Tablespoons olive oil

1 Tablespoon balsamic vinegar or red wine

2 Tablespoons fresh parsley, chopped

Blanch the green beans in boiling water until al dente. Plunge in ice water to cool. Crush the minced garlic with a mortar and pestle if possible. Sauté the garlic in the olive oil briefly in a medium pan until translucent, but not brown. Add the blanched beans and sauté for 5 minutes. Sprinkle with the vinegar or red wine, and garnish with the parsley before serving.

Per serving: calories: 97, protein: 2 gm., carbohydrates: 8 gm, fat: 7 gm.

Minted Green Pea Puree

Serves 3

This will surprise you with its simplicity and sweetness.

1 (10 oz.) package frozen peas

6 mint leaves

1 teaspoon olive oil

salt and pepper to taste

Steam the frozen peas for 2 minutes over boiling water until hot. Be sure not to overcook. Puree the peas with 3 of the mint leaves and the olive oil in a blender. Add salt and pepper to taste, if you wish. Garnish with the remaining mint leaves, and serve.

Optional: Surround with *Curried Mushroom Caps* (see page 35) for a party dish.

Per serving: calories: 93, protein: 4 gm., carbohydrates: 15 gm, fat: 1 gm.

Potatoes and Egg Whites

Serves 4

1½ Tablespoons olive oil

2 cups potatoes, peeled and diced (about 1 lb.)

1 Tablespoon olive oil

6 egg whites

½ teaspoon allspice

¼ teaspoon white pepper

½ teaspoon salt

In a medium, heavy-bottomed skillet, heat the 1½ tablespoons of olive oil over medium heat. Add the diced potatoes, stir to coat with the oil, cover, and steam-fry for 5 minutes. Uncover and fry the potatoes until golden and crispy (about 10-15 minutes), stirring often. Remove the potatoes and set aside. Heat the 1 tablespoon of olive oil in the skillet, add the egg whites, and scramble until firm. Add the fried potatoes, allspice, pepper, and salt, and cook for 2 more minutes. Serve hot.

Per serving: calories: 196, protein: 7 gm., carbohydrates: 23 gm., fat: 8 gm.

Red Peppers and Mushrooms

Serves 4

2 red bell peppers, chopped

2 Tablespoons olive oil

4 cups mushrooms, sliced

½ teaspoon curry powder

5 cloves garlic, chopped

¼ cup pecans, chopped

Sauté the red peppers in the olive oil for 5 minutes over medium heat. Add the mushrooms, curry powder, and garlic, and cook for 5-10 more minutes. Sprinkle with the chopped pecans, and serve right away.

Per serving: calories: 133, protein: 2 gm, carbohydrates: 6 gm, fat: 11 gm.

Sautéed Fennel

Serves 2
Serve as the Italians like it: at room temperature.

1 large fennel bulb, cut in half

1 Tablespoons olive oil

1 clove garlic, minced

2 Tablespoons fresh lemon juice

1 Tablespoon fresh tarragon, chopped

¾ teaspoon salt

pinch of freshly ground black pepper

Sauté the fennel bulb halves and garlic in the olive oil for approximately 3 minutes, so the fennel is still crunchy. Remove from the heat. Add the fresh lemon juice, chopped tarragon, salt and freshly ground pepper. Serve a half bulb per person.

Per serving: calories: 70, protein: 0 gm., carbohydrates: 3 gm., fat: 7 gm.

Baked Peppers

Serves 4

This is easy to do and pretty to look at.

1 yellow bell pepper
1 green bell pepper
1 red bell pepper
1 medium red onion
½ teaspoon salt
1 teaspoon fennel seed
½ teaspoon dried oregano
freshly ground pepper, to taste
1 Tablespoon balsamic vinegar
1 Tablespoon olive oil
2 cloves garlic, minced

Preheat the oven to 400°. Remove the seeds from the bell peppers, and cut the peppers and red onion into 2" wedges. Arrange the vegetables in a 9" x 13" baking pan. Combine the remaining ingredients, except the garlic, and sprinkle them over the vegetables. Bake for 20 minutes. Remove from the oven and sprinkle with the minced garlic.

Per serving: calories: 60, protein: 1 gm., carbohydrates: 6 gm., fat: 2 gm.

Baked Potatoes

Serves 4

4 white potatoes, sliced ¼" thick

⅔ cup low-fat milk

2 cloves garlic, crushed

1 Tablespoon olive oil

6 bay leaves

Place the potatoes in a medium pot, and add the milk and ⅓ cup water. Cook for 20 minutes over medium-high heat. Preheat oven to 400°. Rub an 8" x 8" rectangular baking pan with the crushed garlic and oil. Arrange the bay leaves on the bottom of the pan, and pour in the cooked potatoes and milk. Bake in the oven until brown on top, about 10 minutes.

Per serving: calories: 163, protein: 3 gm., carbohydrates: 29 gm., fat: 4 gm.

Green Pepper and Tomatoes

Serves 4

5 whole medium tomatoes

1½ Tablespoons olive oil

½ large red onion, finely chopped

3 green peppers, finely chopped

4 cloves garlic, minced

1 teaspoon tahini

¼ teaspoon black pepper

Dip the whole tomatoes into a pan of boiling water for 30 seconds. Remove and cool in cold water for a few moments; the skins should slip off easily. Dice and set aside. Heat the olive oil in a large skillet over medium heat. Add the onion and green peppers, and sauté for 10 minutes. Add the garlic, tahini, pepper, and diced tomatoes, cover, and cook for 10 more minutes, stirring occasionally. Serve hot. This is excellent accompanied with rice.

Per serving: calories: 105, protein: 2 gm., carbohydrates: 11 gm., fat: 5 gm.

Delicious Fava Beans

Serves 4

4 cups fresh green fava beans

¼ teaspoon salt

2 bunches coriander

¼ cup olive oil

2 small onions, finely chopped

½ teaspoon black pepper

½ cup fresh lemon juice

3 cloves garlic, finely minced

Simmer the fava beans with the salt and 4 cups of water in a medium pot for 30 minutes, stirring occasionally. Pick the coriander leaves from the stems, and chop the leaves. Heat the olive oil in a large saucepan, and sauté the onions over medium heat for 5 minutes. Add the coriander and sauté for 2 more minutes. Add the cooked beans, black pepper, ¼ cup of the lemon juice, and ¼ cup water to the onions and coriander, and cook for 5 more minutes, stirring occasionally. Pour the bean mixture in a bowl, add the minced garlic and remaining ¼ cup lemon juice, and mix briefly to combine. Refrigerate for 2 hours before serving.

Per serving: calories: 263, protein: 11 gm., carbohydrates: 39 gm., fat: 7 gm.

Fresh Mushrooms and Parsley

Serves 3

1½ Tablespoons olive oil
½ large white onion, finely chopped
2 cloves garlic, minced
2 cups fresh mushrooms, sliced
½ cup water
¼ cup fresh lemon juice
1 cup fresh parsley, chopped

In a medium pot, heat the olive oil over medium heat, add the onions and garlic, and sauté for 5 minutes. Add the mushrooms and sauté for another 5 minutes. Add the water, increase the heat to high, and cook for 5 more minutes. Then add the lemon juice, reduce the heat to low, cover, and cook for 10 more minutes. Sprinkle with the chopped parsley, and chill before serving.

Per serving: calories: 94, protein: 1 gm., carbohydrates: 8 gm., fat: 7 gm.

Sautéed Zucchini and Onions

Serves 3

1½ Tablespoons olive oil
1 large white onion, finely chopped
5 small zucchini, finely chopped
6 garlic cloves, minced
¼ teaspoon salt
⅓ cup fresh lemon juice
2 Tablespoons fresh mint, chopped

Heat the olive oil in a large skillet over medium heat. Add the chopped onion and sauté for 10 minutes. Add the zucchini, stir, and cook uncovered for 10 more minutes, stirring occasionally. Add the garlic, salt, and lemon juice to the zucchini, and stir to combine. Cook for another 10 minutes, sprinkle with the chopped mint, and serve.

Per serving: calories: 91, protein: 2 gm., carbohydrates: 9 gm., fat: 7 gm.

Sautéed Peppers and Tomatoes

Serves 6

8 small whole tomatoes

2 Tablespoons olive oil

4 green peppers, chopped

10 cloves garlic, finely minced

½ teaspoon salt

¾ cup fresh coriander leaves, finely chopped

Drop the whole tomatoes in boiling water for about 30 seconds. Remove and cool in cold water; the skins should slip off easily. Chop and set aside. Heat the olive oil in a large skillet, and sauté the green pepper over medium heat for 10 minutes. Add the chopped tomatoes, cover, and cook for 10 more minutes, stirring occasionally. Add the garlic, salt, and coriander to the tomatoes and peppers, and cook for 5 more minutes. Set aside and serve at room temperature.

Per serving: calories: 90, protein: 1 gm., carbohydrates: 10 gm., fat: 5 gm.

Sautéed Okra with Onion

Serves 4-6

⅓ cup olive oil

4 cups small whole okra

½ large white onion, finely chopped

2 Tablespoons fresh coriander leaves, chopped

1 cup tomato sauce

½ cup water

4 cloves garlic, finely minced

½ teaspoon black pepper

salt to taste

½ cup fresh lemon juice

In a large skillet, heat the olive oil over medium heat, and sauté the whole okra for 5 minutes. Remove the okra from the skillet, and set aside. Add the chopped onion to the skillet, and sauté for 5 minutes. Add the sautéed okra, coriander leaves, tomato sauce, water, garlic, pepper, and salt, and stir to combine. Cover the skillet and cook for 30 minutes. Add the lemon juice 5 minutes before the end of the cooking time.

Per serving: calories: 179, protein: 2 gm., carbohydrates: 12 gm., fat: 13 gm.

Sautéed Peppers and Potatoes

Serves 6

4 large potatoes, peeled and cut into quarters

¼ cup olive oil

1½ red bell peppers, chopped

1½ green peppers, chopped

5 cloves garlic, minced

¼ cup red wine vinegar

¼ cup water

¼ teaspoon black pepper

¼ teaspoon cumin

½ teaspoon anise

Simmer the potatoes in enough water to cover until barely tender. While the potatoes are cooking, heat the olive oil in a large pan, and sauté the peppers over medium-high heat for 10 minutes. Add the garlic, vinegar, water, black pepper, cumin, and anise to the peppers, and stir. Cook for 5 more minutes, stirring occasionally. Drain the cooked potatoes and cut into chunks. Combine the potatoes and sautéed peppers, and serve warm.

Per serving: calories: 172, protein: 1 gm., carbohydrates: 21 gm., fat: 9 gm.

Small Potatoes and Tomatoes

Serves 4-6

8 small white potatoes, sliced 2" thick

6 Roma tomatoes, chopped

3 cloves garlic, crushed

2 carrots, finely chopped

2 stalks celery, finely chopped

1 Tablespoon fresh rosemary, chopped

salt and pepper to taste

2 Tablespoons olive oil

Combine all the ingredients in a medium saucepan. Cover and cook over low heat for 30 minutes, until the tomato juices have evaporated and the potatoes are tender.

Per serving: calories: 277, protein: 3 gm., carbohydrates: 49 gm., fat: 4 gm.

Tuscan Baked Vegetables

Serves 6

A wonderfully easy method for cooking vegetables that Marilyn learned about in Tuscany. This is a dish delicious enough to serve to company. It is important to serve these vegetables at room temperature, as the refrigerator robs them of their flavor. This makes this an ideal item for buffets and cook-ahead meals.

2 red bell peppers

1 yellow bell pepper

1 medium eggplant

1 medium red onion

2 Tablespoons olive oil

3 Tablespoons Parmesan cheese

¼ cup water

3 Tablespoons balsamic vinegar or red wine vinegar

1 teaspoon dried oregano

Preheat the oven to 350°. Cut vegetables into 2½" wedges. Place in a 9" x 13" baking pan, brush with the olive oil, and sprinkle with the Parmesan cheese. Pour the water into the bottom of the pan, and bake for 20 minutes. Sprinkle with the balsamic or red wine vinegar and oregano, and serve.

Per serving: calories: 95, protein: 2 gm., carbohydrates: 10 gm., fat: 5 gm.

Bean Salad with Tofu and Rosemary

Serves 4

Excellent for buffets or luncheons.

1 (16 oz.) can cooked beans

1 medium red onion, finely chopped,

2 Tablespoons fresh rosemary, finely chopped, or 1 Tablespoon dried rosemary, crumbled

½ lb. firm tofu, crumbled

1 Tablespoon white miso mixed with 2 Tablespoons water

2 teaspoons kelp powder (opt.)

2 Tablespoons olive oil

3 Tablespoons red wine vinegar

2 cloves garlic, minced

salt and freshly ground black pepper to taste

1 cup fresh parsley, chopped

Combine the cooked beans, onion, rosemary, crumbled tofu, miso and water, and kelp powder (if using), in a large salad bowl. Mix the oil, vinegar, garlic, salt, and pepper in a small bottle, and pour over the salad. Toss to blend the flavors, garnish with the parsley, and serve at room temperature.

Per serving: calories: 292, protein: 15 gm., carbohydrates: 36 gm., fat: 8 gm.

Green Beans and Tomatoes

Serves 7

⅓ cup olive oil

2 medium white onions, chopped

5 cloves garlic, sliced

8 cups fresh green beans, chopped (about 2 lbs.)

2 (14.5 oz.) cans whole tomatoes, peeled

1 cup water

½ teaspoon salt

½ teaspoon black pepper

1 cup fresh mint, chopped

In a large pot, heat the olive oil and sauté the onions and garlic for 5 minutes over medium-high heat. Add the green beans and sauté for 5 more minutes over medium heat. Add the tomatoes, water, salt, and pepper; cover, and cook for 30 minutes. Cool to room temperature. Before serving, top with the fresh mint.

Per serving: calories: 156, protein: 3 gm., carbohydrates: 15 gm., fat: 9 gm.

Vegetable Casserole

Serves 8

2 medium onions
1 small eggplant
3 medium potatoes
⅓ cup olive oil
2 small zucchini, sliced
1 green pepper, chopped
2 carrots, sliced
8 cloves garlic, chopped
3 cups water
2 (15 oz.) cans tomato sauce
½ teaspoon cinnamon
½ teaspoon salt

Peel the onions, eggplant, and potatoes, and cut into 2" cubes. Heat 3 tablespoons of the olive oil in a large pot, and sauté the onions for 5 minutes over high heat. Add the remaining olive oil, the eggplant, potatoes, zucchini, green pepper, carrots, and garlic to the onions, and sauté for 5 more minutes. Add the water, tomato sauce, cinnamon, and salt to the vegetables. Bring to a boil, reduce the heat, cover, and cook for 40 minutes over medium heat.

Per serving: calories: 189, protein: 2 gm., carbohydrates: 26 gm., fat: 8 gm.

Moussaka

Makes 3 servings

1 large eggplant, peeled and sliced ½" thick

1½ Tablespoons olive oil

¼ cup textured vegetable protein granules

1 medium yellow onion, chopped

2 cloves garlic, chopped

2 tomatoes, chopped

1 Tablespoon dried oregano

½ teaspoon cinnamon

¼ teaspoon salt

¼ teaspoon cumin

¼ cup lemon juice

Place the eggplant slices on a cookie sheet, and brush with half of the olive oil. Broil until brown, watching carefully not to burn it. Reconstitute the textured vegetable protein granules in ¼ cup boiling water, and set aside. Sauté the onion in the remaining olive oil until well browned, stirring frequently. Add the garlic for the last 2 minutes. Add the chopped tomatoes, oregano, cinnamon, salt, cumin, the reconstituted textured vegetable protein, and lemon juice. Cook for an additional 10 minutes to blend flavors. Preheat the oven to 350°. Assemble in a medium baking dish by alternating layers of eggplant and the above sauce. Bake 25 minutes and serve hot.

Per serving: calories: 163, protein: 5 gm., carbohydrates: 21 gm., fat: 7 gm.

Pepper Ragout

Makes 4 servings

1 Tablespoon olive oil

½ large yellow onion, chopped

2 shallots, chopped

4 bell peppers, seeded and sliced

2½ cups tomatoes, chopped

1 Tablespoon tomato paste

½ cup textured vegetable protein granules

½ cup black olives, chopped

¼ cup white wine

1 Tablespoon dried basil

salt and pepper to taste

¾ cup low-fat soymilk

1 Tablespoon white wine

1 Tablespoon ground almonds

½ teaspoon Dijon mustard

2 Tablespoons flour

2 Tablespoons soy Parmesan cheese

In a large sauté pan, heat the olive oil over medium heat. Add the onion and shallots, and saute for 5 minutes. Add the peppers, tomatoes, tomato paste, textured vegetable protein, olives, ¼ cup white wine, basil, salt, and pepper; mix well and simmer until almost dry. Pour into a medium baking dish. Make a sauce by combining the soymilk, 1 Tablespoon white wine, almonds, mustard, and flour in a small saucepan. Bring to a boil over medium heat, stirring constantly until thick. Cover the vegetables with the sauce, sprinkle with the soy Parmesan cheese, and put under a broiler until brown.

Per serving: calories: 219, protein: 10 gm., carbohydrates: 21 gm., fat: 9 gm.

Salads

Thyme and Potato Salad

Serves 5

5 medium potatoes
¼ cup fresh thyme, chopped
½ medium white onion, minced
¼ cup extra virgin olive oil
¼ teaspoon salt
⅓ cup fresh lemon juice
1 teaspoon balsamic vinegar

Boil the potatoes until tender. While they are cooking, wash the thyme and pick off the leaves. Add the minced onion to the thyme. Drain the cooked potatoes, peel, dice, and add to the onions and thyme. Mix the olive oil, salt, lemon juice, and vinegar in a small jar, and pour over the potatoes, onions, and thyme. Toss well.

Per serving: calories: 221, protein: 2 gm., carbohydrates: 29 gm., fat: 10 gm.

French Tomato Salad

Serves 6

6 medium tomatoes, finely chopped

1 cucumber, finely sliced

½ small white onion, peeled and finely chopped

3 Tablespoons fresh mint, finely chopped

2 Tablespoons extra virgin olive oil

½ cup fresh lemon juice

1 Tablespoon Dijon mustard

freshly ground black pepper to taste

Toss the tomatoes, cucumber, onion, and mint in a large bowl. Mix the olive oil, lemon juice, and mustard for a dressing. Add the dressing to the salad, and toss. Sprinkle with freshly ground black pepper, and serve right away.

Per serving: calories: 82, protein: 1 gm., carbohydrates: 8 gm., fat: 5 gm.

Holiday Salad

Serves 6

This festive salad is fun to arrange, . . . and delicious.

6 oranges

4-5 endive leaves

1 bunch watercress

3 Tablespoons balsamic vinegar

3 Tablespoons fresh lemon juice

1 teaspoon Dijon mustard

⅓ cup olive oil

salt and pepper to taste

Peel the oranges, leaving them whole, and slice into ⅓" rounds. Place the slices in the center of a round serving dish, mounding them up higher in middle. Arrange the endive leaves around the oranges with tips pointing out like a flower. Place watercress around the edge of the plate like a border. Prepare a vinaigrette dressing by combining the vinegar, lemon juice, and mustard, then adding the oil, salt, and pepper. Pour over the salad and serve.

Per serving: calories: 175, protein: 2 gm., carbohydrates: 16 gm., fat: 10 gm.

Italian Carrot Salad

Serves 4

This salad is served frequently in Italy.

6 medium carrots, sliced in long, thin, matchstick pieces

¼ cup currants

½ cup fresh parsley, chopped

1 teaspoon Dijon mustard

1 Tablespoon balsamic vinegar

3 Tablespoons lemon juice

1 teaspoon salt

2 Tablespoons extra virgin olive oil

2 Tablespoons water

Put the carrot pieces, currants, and parsley in large salad bowl. Mix together the mustard, vinegar, lemon juice, and salt until blended. Add the olive oil and water, and stir well. Pour over the salad and toss well.

Per serving: calories: 125, protein: 1 gm., carbohydrates: 15 gm., fat: 6 gm.

Minted Fruit Salad

Serves 4-6

4 navel oranges

1 cantaloupe, peeled and sliced in bite size pieces

1 pint fresh strawberries

½ cup fruity white wine or ginger ale

1 teaspoon extra virgin olive oil

1 Tablespoon honey (optional)

3 Tablespoons fresh mint, chopped

1 Tablespoon freshly grated ginger

Peel and slice the oranges and cantaloupe over a small bowl to catch the juice; reserve the juice. Put the oranges and strawberries in a large serving bowl (a glass bowl shows off the colorful fruit). Add the reserved juice to the remaining ingredients, and toss well.

Per serving: calories: 152, protein: 2 gm., carbohydrates: 26 gm., fat: 1 gm.

Mixed Vegetable Salad

Serves 10

½ cup extra virgin olive oil

¼ cup water

6 cloves garlic, finely minced

½ cup fresh lemon juice

¼ cup red vinegar

½ teaspoon black pepper

1 romaine lettuce, chopped

6 small tomatoes, chopped

1 cucumber, peeled and sliced

6 radishes, chopped

1 cup fresh mint

1 cup fresh parsley, finely chopped

1 large white onion, chopped

To make a dressing, mix the olive oil, water, garlic, lemon juice, vinegar, and black pepper in a pint jar. In a large bowl, mix the lettuce, tomatoes, cucumbers, radishes, mint, parsley, and onion. Add the dressing and mix. Set aside for 30 minutes and serve.

Per serving: calories: 134, protein: 1 gm., carbohydrates: 8 gm., fat: 11 gm.

Orange and Fennel Salad

Serves 4

Fennel makes a delightful change from cucumbers. Italians love it and use it often. The feathery tops are good too and are pretty in a salad.

3 oranges

1 large fennel bulb, sliced, or 2 medium cucumbers, seeded and sliced

3 Tablespoons green onions, finely chopped

⅓ cup Italian Dressing, page 128

1 bunch watercress

Peel the oranges, leaving them whole, and cut into slices. Toss the orange slices, cucumbers, and onions with the Italian dressing. Arrange on a bed of watercress. Serve chilled.

Per serving: calories: 169, protein: 1 gm., carbohydrates: 15 gm., fat: 11 gm.

Greek Romaine Salad

Serves 6

½ head romaine lettuce, chopped

5 small tomatoes, chopped

⅓ cup fresh lemon juice

1 clove garlic, minced

½ teaspoon black pepper

2 Tablespoons extra virgin olive oil

10 Greek olives, finely chopped

3 Tablespoons feta cheese, crumbled

In a large bowl, toss the lettuce and tomatoes. To make a dressing, mix the lemon juice, garlic, black pepper, and olive oil in a small jar. Add the dressing to the lettuce and tomatoes, and toss lightly to combine. Top with the chopped olives and crumbled feta cheese.

Per serving: calories: 93, protein: 2 gm., carbohydrates: 5 gm., fat: 7 gm.

Thyme Salad

Serves 3

¼ cup fresh thyme

¼ medium white onion, very finely chopped

3 large tomatoes, finely chopped

½ teaspoon Dijon mustard

1½ Tablespoons extra virgin olive oil

Combine all the ingredients in a medium bowl, set aside for 30 minutes so the flavors will blend, and serve.

Per serving: calories: 86, protein: 1 gm., carbohydrates: 6 gm., fat: 7 gm.

Spinach Salad

Serves 4

1 lb. spinach

5 small tomatoes, chopped

¼ large white onion, finely chopped

½ cup fresh lemon juice

½ teaspoon freshly ground black pepper

3 Tablespoons extra virgin olive oil

Wash the spinach, pick the leaves from the stems, discard the stems, and chop the leaves. In a large bowl, mix the spinach with the tomatoes and onions. To make a dressing, combine the lemon juice, black pepper, and olive oil in a small jar. Add the dressing to the salad, mix for 1 minute, and serve.

Per serving: calories: 155, protein: 3 gm., carbohydrates: 12 gm., fat: 10 gm.

Tomato Salad

Serves 5

4 medium tomatoes, chopped
1 medium cucumber, chopped
½ small white onion, finely chopped
¼ cup extra virgin olive oil
1 clove garlic, finely minced
½ teaspoon black pepper
⅓ cup fresh lemon juice

Mix the tomatoes, cucumber, and onion in a medium bowl. Combine the olive oil, garlic, black pepper, and lemon juice in a small jar. Pour over, mix, and serve.

Per serving: calories: 129, protein: 1 gm., carbohydrates: 7 gm., fat: 10 gm.

Orange Salad

Serves 6

3 oranges, peeled and chopped
8 black olives, sliced
1 large head romaine lettuce
⅓ cup fresh orange juice
1 Tablespoon red vinegar
¼ teaspoon paprika
¼ cup extra virgin olive oil

Toss the oranges, olives, and lettuce in a salad bowl. Combine the remaining ingredients for a dressing. Pour over the dressing, toss lightly, and serve.

Per serving: calories: 141, protein: 1 gm., carbohydrates: 10 gm., fat: 11 gm.

Artichoke Salad

Serves 4

A lovely luncheon dish.

¼ cup walnuts

1 (10 oz.) package frozen artichoke hearts, thinly sliced

2 cloves garlic, minced

1 fennel bulb, thinly sliced

1 Tablespoon olive oil

1 (11 oz.) can mandarin oranges

1 lb. fresh spinach, chopped

¼ cup Raspberry Vinaigrette **(see page 127)**

Dry-roast the walnuts in a heavy-bottomed pan or on a baking sheet in a 350° oven (watch carefully until slightly browned). Sauté the artichoke hearts, garlic, and fennel in the olive oil. Toss all the ingredients with the Raspberry Vinaigrette, and serve.

Per serving: calories: 259, protein: 4 gm., carbohydrates: 23 gm., fat: 16 gm.

Zucchini and Red Pepper Salad

Serves 5

8 cloves garlic, peeled, smashed, and minced

1 Tablespoon olive oil

1½ pound zucchini, cut in matchsticks

salt and black pepper, to taste

2 medium red peppers, julienned

½ cup green olives, sliced

1 teaspoon thyme

½ cup onions, chopped

1 recipe Caper Dressing (see page 122)

Briefly heat the garlic in the olive oil over medium heat in a large frying pan for 1 minute. Add the zucchini, salt and black pepper, and red peppers, and sauté for 5 minutes. Remove from the heat. Add the olives, thyme, and onions, and put in a serving bowl. Pour over the Caper Dressing, chill 3 hours, and serve.

Per serving: calories: 168, protein: 1 gm., carbohydrates: 8 gm., fat: 13 gm.

Hearty Broccoli Salad

Serves 4

½ cup textured vegetable protein chunks

1 Tablespoon olive oil

3 cups broccoli florets

¼ cup unsalted peanuts

¼ cup lemon juice

1 Tablespoon red vinegar

Reconstitute the textured vegetable protein chunks in ½ cup boiling water for 10 minutes; drain. Heat the olive oil in a large skillet, and sauté the textured vegetable protein over medium heat for 5 minutes. Add the broccoli and peanuts, and cook for 10 more minutes. Add the lemon juice and vinegar 2 minutes before the end of the cooking time, and stir.

Per serving: calories: 139, protein: 8 gm., carbohydrates: 9 gm., fat: 8 gm.

Green Bean Salad

Serves 8

8 cups fresh green beans, cut in 2" pieces (about 2 lbs.)

6 cloves garlic, finely minced

½ cup fresh lemon juice

¼ cup extra virgin olive oil

Bring 7 cups of water to a boil. Add the green beans, reduce heat to medium, cover, and cook until tender (about 30 minutes). Make a dressing by mixing the garlic, lemon juice, and olive oil. Drain the beans and set aside to cool. Combine the dressing and the beans, and serve cold.

Per serving: calories: 102, protein: 2 gm., carbohydrates: 9 gm., fat: 7 gm.

Lebanese Tabouli

Serves 8

Tabouli has many variations. When the recipe traveled across the Atlantic to this country, it was often made with more bulgur than it is in Lebanon. This rich and hearty version is made with lots of fresh parsley.

1 cup bulgur, washed and drained

1 cup fresh lemon juice

1½ teaspoons salt

2 Tablespoons allspice

½ medium white onion, finely chopped

9 medium tomatoes, chopped

5 cups fresh parsley, finely chopped

1 cup fresh green mint leaves, finely chopped

¼ cup extra-virgin olive oil

Combine 1 cup boiling water to the bulgur in a small bowl; cover and let sit for 15 minutes. Add the lemon juice to the bulgur, and set aside for 5 minutes. Add the remaining ingredients except the olive oil to the bulgur, and mix. Pour the olive oil over all, and mix thoroughly.

Per serving: calories: 181, protein: 4 gm., carbohydrates: 25 gm., fat: 7 gm.

Mushroom and Fennel Salad

Serves 4

1 teaspoon Dijon mustard

3 Tablespoons balsamic vinegar

2 Tablespoons fresh lemon juice

2 Tablespoons olive oil

1 cup mushrooms, thinly sliced

1 medium red onion, thinly sliced

1 large fennel bulb, cut in half and sliced ⅓" thick

2 Tablespoons chives, chopped

Blend the mustard, vinegar, and lemon juice well, then add the oil. Marinate the mushrooms, red onion, and fennel for 20 minutes in the dressing. Add chives, toss, and serve.

Per serving: calories: 88, protein: 1 gm., carbohydrates: 5 gm., fat: 7 gm.

Potato and Coriander Salad

Serves 4

4 medium potatoes

3 cloves garlic, minced

¼ teaspoon salt

⅓ cup fresh lemon juice

2 Tablespoons extra virgin olive oil

2 Tablespoons water

½ cup fresh coriander, chopped

Boil the whole potatoes, slip off the skins, and dice. Combine the garlic, salt, lemon juice, olive oil, and water, and mix for 30 seconds. Pour over the potatoes and toss. Add the coriander to the potatoes, and mix for 1 minute. Serve cold.

Per serving: calories: 177, protein: 2 gm., carbohydrates: 29 gm., fat: 7 gm.

Beet Salad

Serves 6

½ large, white onion, finely chopped

6 beets, boiled and chopped

1 cup fresh parsley, finely chopped

¼ cup extra virgin olive oil

¼ cup red vinegar

½ cup fresh lemon juice

Mix the onion, beets, and parsley, and mix very well. Combine the olive oil, vinegar, and lemon juice, and pour over the beets. Chill and serve.

Per serving: calories: 111, protein: 1 gm., carbohydrates: 7 gm., fat: 9 gm.

Potato and Egg Salad

Serves 8

¼ cup fresh lemon juice
⅓ cup white vinegar
1 teaspoon allspice
1 teaspoon Dijon mustard
⅓ cup extra virgin olive oil
8 potatoes, boiled and diced
6 egg whites, hard boiled and diced
¼ large white onion, chopped
½ cup fresh parsley, chopped

To make a dressing, combine the lemon juice, vinegar, allspice, mustard, and olive oil. Toss together the diced potatoes, egg whites, onions, and parsley. Add the dressing and mix.

Per serving: calories: 209, protein: 4 gm., carbohydrates: 29 gm., fat: 8 gm.

Potato and Mint Salad

Serves 5

5 medium potatoes

¼ teaspoon salt

1 teaspoon ground cumin

2½ Tablespoons extra virgin olive oil

¼ medium white onion, finely chopped

1½ cup fresh mint, chopped

Boil the potatoes until tender, then peel and dice them. Combine the salt, cumin, and olive oil, then add to the potatoes, along with the onion and mint. Toss thoroughly.

Per serving: calories: 175, protein: 2 gm., carbohydrates: 28 gm., fat: 7 gm.

Tomato and Basil Salad

Serves 3

1 cup fresh basil leaves

6 small tomatoes, minced

3 teaspoons red vinegar

⅓ cup fresh lemon juice

2 Tablespoons extra virgin olive oil

¼ teaspoon salt

Wash the basil, pick the leaves from stems, discard the stems, and chop the leaves finely. Place the minced tomatoes on a plate, and cover with the chopped basil. To make a dressing, mix the vinegar, lemon juice, olive oil, and salt in a small jar. Pour over the tomatoes and basil. Chill for one hour and serve.

Per serving: calories: 135, protein: 2 gm., carbohydrates: 12 gm., fat: 9 gm.

Beautiful Summer Vegetable Salad

Serves 5

A good salad to serve with a homemade soup.

½ cup Vinaigrette Dressing (see page 133)

1 medium red onion, sliced thin

1 (7 oz.) jar roasted red peppers, cut in strips

¾ cup croutons

1 clove garlic, minced

½ Tablespoon olive oil

¼ pound green beans, cut in 2" pieces

3 ripe tomatoes, sliced in quarters

½ pound feta or goat cheese, rinsed and cubed (opt.)

1 cup basil leaves

assortment of salad greens of your choice (arugula, radicchio, red oak leaf, etc.) equal to one head

Pour the vinaigrette over the onions and peppers in a large salad bowl, and marinate for 2 hours. Meanwhile, sauté the croutons in the garlic and olive oil. Layer the green beans, tomatoes, feta or goat cheese, basil, and greens on top of the marinated vegetables. Top with the croutons, toss, and serve.

Per serving: calories: 166, protein: 2 gm., carbohydrates: 12 gm., fat: 11 gm.

Beet Salad with Tofu Vinaigrette

Makes 4 servings

4 fresh whole beets

2 green apples, sliced 1" thick

3 Tablespoons lemon juice

½ pound mixed baby salad greens

½ cup Tofu Vinaigrette (see page 140)

¼ cup whole hazelnuts

Place the beets in a small saucepan, and half-cover with water. Cover and simmer for 30 minutes or until tender. Once done, remove the skins and cut into 1" slices. Arrange the beets and apples over the baby greens. Pour the lemon juice over, and let sit at room temperature for 10 minutes. Top with Tofu Vinaigrette and whole hazelnuts, and serve.

Per serving: calories: 197, protein: 3 gm., carbohydrates: 18 gm., fat: 11 gm.

Hot Mushroom Tofu Salad

Makes 4 servings

1 pound mixed mushrooms (oyster, shiitake), chopped

¼ cup olive oil

½ cup soft tofu, drained, dried and cubed

1 large shallot, minced

3 garlic cloves, minced

¼ cup balsamic vinegar

2 Tablespoons basil, chopped

1 Tablespoon dried oregano

2 Tablespoons lemon juice

freshly ground pepper to taste

1 pound mixed baby salad greens

Sear the mushrooms in 3 tablespoons of the olive oil in a large, very hot sauté pan. Stir after 30 seconds and continue stirring for 2½ minutes longer until nicely browned. Add the tofu, shallots, garlic, and the remaining olive oil, and continue sautéing for 30 more seconds. Pour over the balsamic vinegar, and scrape to deglaze the pan. Add the herbs, lemon juice, and pepper. Pour this hot dressing over the baby salad greens, toss well, and serve immediately.

Per serving: calories: 268, protein: 6 gm., carbohydrates: 26 gm., fat: 14 gm.

Tofu Salad

Serves 5

1 Tablespoon sesame oil

1 Tablespoon extra virgin olive oil

3 Tablespoons soy sauce

¼ teaspoon crushed red pepper flakes

1 teaspoon sugar

1 stalk celery, sliced

2 carrots, peeled and thinly sliced

1 cup soft tofu, drained, dried and thinly sliced

8 dried porcini mushrooms, stemmed and thinly sliced (opt.)

In a medium bowl, combine the sesame oil, olive oil, soy sauce, red pepper, and sugar. Add the celery, carrots, tofu, and mushrooms, and toss with the dressing. Chill for 3 hours before serving.

Per serving: calories: 121, protein: 5 gm., carbohydrates: 9 gm., fat: 8 gm.

Peanut Tofu Salad

Makes 4 servings

1 Tablespoon sugar

1 Tablespoon extra virgin olive oil

1½ Tablespoons soy sauce

3 Tablespoons white wine vinegar

1 teaspoon sesame oil

¼ teaspoon ground red pepper

½ teaspoon salt

1 cup firm tofu, drained and cubed

2 carrots, shredded

2 cups bean sprouts

2 green onions, thinly sliced

½ cup salted peanuts, chopped

In a small bowl, mix the sugar, olive oil, soy sauce, vinegar, sesame oil, pepper, and salt. In a large salad bowl, place the tofu, carrots, bean sprouts, onions, and peanuts; toss gently. Add the dressing and toss one more time to combine.

Per serving: calories: 245, protein: 11 gm., carbohydrates: 15 gm., fat: 15 gm.

Low-Cal Couscous Salad

Makes 6 servings

¼ cup textured vegetable protein granules

2 cups uncooked couscous

1 red pepper, finely chopped

1 green pepper, finely chopped

1 bunch scallions, finely chopped (including greens)

⅓ cup cilantro, chopped

1 Tablespoon extra virgin olive oil

1 Tablespoon water

½ cup red wine vinegar

1 teaspoon dried thyme, or 1 Tablespoon fresh thyme, minced

In a large salad bowl, combine the textured vegetable protein to the couscous, and mix well. Bring 2½ cups of water to a boil; add the water to the couscous mixture, stir, and cover for 15 minutes. Fluff with a fork and add the peppers, scallions, and cilantro. Prepare a vinaigrette by combining the olive oil, water, wine vinegar, and thyme; pour over the salad and mix well. Pack the salad into small, oiled molds, and chill for 2 hours. Turn out on salad plates, and serve.

Per serving: calories: 165, protein: 6 gm., carbohydrates: 30 gm., fat: 2 gm.

Indian Stuffed Tomatoes

Makes 5 servings

10 large firm but ripe tomatoes, scooped out (reserve pulp)

½ cup textured vegetable protein granules

1 green pepper, finely chopped

3 Tablespoons capers

1 dill pickle, finely chopped

⅓ cup olive oil

⅓ cup balsamic vinegar

¼ cup lemon juice

4 cups cooked rice

½ Tablespoon curry powder

¼ cup cilantro, chopped

Chill the tomato shells until ready to fill. Reconstitute the textured vegetable protein granules in ½ cup boiling water; set aside. Puree the tomato pulps in a blender. Place in a medium bowl, add the green pepper, capers, pickle, olive oil, vinegar, lemon juice, and mix. Add the vegetables and textured vegetable protein to the rice, and mix well. Chill for 4 hours. Once chilled, fill the tomatoes with the rice mixture. Sprinkle the cilantro on top, and serve.

Per serving: calories: 389, protein: 10 gm., carbohydrates: 54 gm., fat: 14 gm.

Sauces, Dressings, and Marinades

Basil and Mustard Dressing

Makes 1 ½ cups of dressing

A delightful dressing used mostly with fresh green salads.

1 ½ Tablespoons Dijon mustard

¼ cup fresh lemon juice

3 Tablespoons red wine vinegar

½ cup water

½ cup fresh basil, chopped

¼ teaspoon salt

½ cup extra virgin olive oil

Combine all the ingredients well in a blender. Refrigerate for 1 hour before serving.

Per Tbsp.: calories: 42, protein: 0 gm., carbohydrates: 0 gm., fat: 4 gm.

Caper Dressing

Makes ⅔ cup

3 Tablespoons capers, well rinsed and drained

3 Tablespoons red wine or white vinegar

½ teaspoon oregano

½ teaspoon sugar

¼ cup extra virgin olive oil

Combine all the ingredients together thoroughly right before serving.

Per Tbsp.: calories: 50, protein: 0 gm., carbohydrates: 0 gm., fat: 5 gm.

Caper Pine Nut Sauce

Makes approximately 1¼ cup sauce. Excellent on cooked vegetables.

¼ cup pine nuts

¼ cup capers, drained and well rinsed

2½ Tablespoons red wine vinegar

1 slice whole wheat bread, crusts removed

1 teaspoon dried oregano

¼ cup olive oil

¼ cup water

Puree all the ingredients in a blender.

Per 2 Tbsp.: calories: 74, protein: 1 gm., carbohydrates: 2 gm., fat: 7 gm.

Balsamic-Lemon Vinaigrette

Makes 1 cup

¼ cup fresh thyme

1 teaspoon freshly ground black pepper

3 cloves garlic, minced

1 Tablespoon Dijon mustard

2 shallots, minced

⅓ cup balsamic vinegar

3 Tablespoon fresh lemon juice

¼ cup extra virgin olive oil

zest of one lemon

salt to taste

Combine the thyme, pepper, garlic, mustard, and shallots in a food processor. Add the balsamic vinegar, lemon juice, olive oil, and lemon zest; mix well. Add salt as desired and serve right away.

Per Tbsp.: calories: 34, protein: 0 gm., carbohydrates: 1 gm., fat: 3 gm.

Middle Eastern Sauce

Makes 1⅔ -¾ cups

This delicious sauce is used in almost every fishing village in the Middle-East and is known as tarator. *You can use it with any vegetarian sandwich or as a side dip.*

1 cup fresh parsley, finely chopped

¼ cup tahini

4 cloves garlic, finely minced

⅓ cup lemon juice

¼ cup water

1 Tablespoon extra virgin olive oil

Combine all the above ingredients in a blender for 15 seconds. Serve right away.

Per Tbsp.: calories: 35, protein: 1 gm., carbohydrates: 2 gm., fat: 2 gm.

Garlic Sauce

Makes ¼ cup

This sauce is super hot and only for real garlic lovers! It makes a great sandwich spread.

8 cloves garlic, finely crushed and minced

¼ teaspoon salt

1½ Tablespoons extra virgin olive oil

2 Tablespoons fresh lemon juice

Add the olive oil and salt to the garlic, and let stand for 1 minute. Add the lemon juice to the garlic mixture, and whip with a whisk or beater until creamy.

Per Tbsp.: calories: 55, protein: 0 gm., carbohydrates: 3 gm., fat: 5 gm.

Rice Sauce

Makes 1½ cups of sauce

If you can get Italian parsley, it adds a distinctive flavor.

¼ cup olive oil

1 whole head garlic, minced

1½ cups fresh regular or Italian parsley leaves, finely chopped

2 medium tomatoes, finely chopped

¼ teaspoon freshly ground black pepper

salt to taste

Heat the olive oil in a small saucepan, and sauté the garlic over medium heat for 4 minutes. Add the parsley, sauté for 2 more minutes, then add the tomatoes and black pepper, and cook for 5 more minutes. Serve hot.

Per ¼ cup: calories: 98, protein: 1 gm., carbohydrates: 4 gm., fat: 9 gm.

Green Sauce

Makes ¾ cup

This sauce can be refrigerated for several days.

3 Tablespoons fresh parsley, chopped

1 teaspoon garlic, chopped coarsely

½ teaspoon Dijon mustard

1 tablespoon fresh lemon juice

½ cup extra virgin olive oil

Puree all ingredients in a blender. Taste to adjust flavor.

Per Tbsp.: calories: 80, protein: 0 gm., carbohydrates: 0 gm., fat: 9 gm.

Parsley and Yogurt Sauce

Makes 2½ cups

Use with fish, sandwiches, and vegetables.

2 cups non-fat, plain yogurt

½ cup fresh parsley, finely chopped

2 cloves garlic, minced

2 Tablespoons fresh lemon juice

1 Tablespoon extra virgin olive oil

Combine all the ingredients in a blender for 15 seconds. Cover and refrigerate at least 2 hours before serving.

Per ¼ cup: calories: 42, protein: 3 gm., carbohydrates: 4 gm., fat: 1 gm.

Pesto Sauce

Makes 1⅔ cup

Use with any hot pasta dish, omelettes, or rice.

1 cup fresh basil

¼ cup fresh parsley, chopped

3 Tablespoons Parmesan cheese, grated

3 cloves garlic, finely minced

3 Tablespoons pistachios

½ cup hot water

¼ teaspoon salt

¼ cup extra virgin olive oil

Combine all the ingredients in a blender or food processor at medium speed until smooth.

Per ¼ cup: calories: 112, protein: 2 gm., carbohydrates: 2 gm., fat: 11 gm.

Raspberry Vinaigrette

Makes approximately ½ cup

¼ cup raspberry vinegar

2 cloves garlic, peeled, smashed, and finely chopped

⅓ cup extra virgin olive oil

Combine all the ingredients in a small jar.

Per Tbsp.: calories: 76, protein: 0 gm., carbohydrates: 1 gm., fat: 7 gm.

Italian Dressing

Makes approximately ¾ cup

1¼ Tablespoons white wine vinegar

2 Tablespoons fresh parsley, finely chopped

1 clove garlic or shallot, finely chopped

1 teaspoon oregano

¼ teaspoon freshly ground pepper

1 Tablespoon fresh lemon juice

½ cup extra virgin olive oil

Combine all the ingredients in a small jar.

Per Tbsp.: calories: 80, protein: 0 gm., carbohydrates: 0 gm., fat: 9 gm.

Italian Piquant Dressing

Makes 2 cups

1 cup low-fat, plain yogurt
1 Tablespoon fresh lemon juice
2 Tablespoons vinegar
1 (16 oz.) jar of marinated Italian vegetables, rinsed and drained
1 teaspoon dried oregano
½ cup extra virgin olive oil

Process all the ingredients in the blender until smooth.

Per Tbsp.: calories: 51, protein: 1 gm., carbohydrates: 2 gm., fat: 4 gm.

Thyme Dressing

Makes approximately 1 cup

Use this dressing with salads or as an overnight marinade.

2 teaspoons red vinegar
½ cup fresh lemon juice
2 cloves garlic, minced
2 Tablespoons water
¼ cup extra virgin olive oil
2 Tablespoons fresh thyme

Combine all the ingredients in a small jar, and chill for 2 hours before serving.

Per Tbsp.: calories: 32, protein: 0 gm., carbohydrates: 1 gm., fat: 3 gm.

Marilyn's Marinade

Makes 1 cup

Use any combination of fresh herbs, such as rosemary, basil, thyme, oregano, or parsley. This is delicious on any vegetable, especially cold, cooked vegetables, raw carrots and other crudites, and cooked red potatoes.

½ cup fresh lemon juice

¼ cup Dijon mustard

¼ cup fresh herbs, finely chopped

salt and freshly ground black pepper, to taste

1 Tablespoon extra virgin olive oil

Combine the marinade ingredients in a jar or non-aluminum bowl.

Per Tbsp.: calories: 15, protein: 0 gm., carbohydrates: 1 gm., fat: 1 gm.

Tomato Basil Vinaigrette

Makes approximately 1 ½ cups

¼ cup red wine vinegar

1 Tablespoon Dijon mustard

¾ cup extra virgin olive oil

1 medium tomato, finely chopped and seeded

2 teaspoons shallot, minced

1 jalapeño pepper, seeded and finely chopped

2 Tablespoons fresh basil, cut in small pieces

salt and freshly ground black pepper, to taste

⅛ teaspoon Tabasco sauce

Whisk the vinegar into the mustard to dissolve. Add the oil and blend thoroughly. Mix in the tomato, shallots, jalapeño pepper, basil, salt, black pepper, and Tabasco (handle the jalapeño carefully—wash hands thoroughly or use plastic gloves). Combine well.

Per Tbsp.: calories: 63, protein: 0 gm., carbohydrates: 0 gm., fat: 7 gm.

Tahini Sauce

Makes 1½ cups

Use with sandwiches and falafel.

1 cup tahini

¼ cup water

¼ cup fresh lemon juice

1 teaspoon red vinegar

2 clove garlic, minced

¼ teaspoon salt

1 teaspoon olive oil

Combine all the ingredients in a bowl, and mix for about 2 minutes.

Per Tbsp.: calories: 66, protein: 2 gm., carbohydrates: 3 gm., fat: 5 gm.

Yogurt and Cucumber Dressing

Makes 2 cups

Use this dressing with rice as well as salads.

1 cup non-fat, plain yogurt

½ cucumber, peeled and sliced

6 fresh mint, chopped

2 cloves garlic, finely minced

1 teaspoon extra virgin olive oil

Combine all the ingredients in a blender, and pulse for 15 seconds. Refrigerate at least 2 hours before serving.

Per ¼ cup: calories: 25, protein: 2 gm., carbohydrates: 2 gm., fat: 0 gm.

Vinaigrette Dressing

Makes ¾ cup

- **3 cloves garlic, peeled, and minced, or 2 shallots, minced**
- **2 Tablespoons parsley, finely chopped**
- **1 teaspoon sea salt**
- **pinch of freshly ground pepper**
- **2 Tablespoons red wine vinegar**
- **3 Tablespoons lemon**
- **1 teaspoon Dijon mustard**
- **⅓ cup extra virgin olive oil**

Smash the garlic, parsley, salt, and pepper with a mortar and pestle, if you have one, or mix in a blender. Add the vinegar, lemon, and mustard, and mix well. Stir in the oil.

Per Tbsp.: calories: 53, protein: 0 gm., carbohydrates: 1 gm., fat: 5 gm.

Olive Sauce

Makes 2 cups

This tasty sauce can be used with pasta dishes.

20 Kalamata olives, rinsed and pitted

2 shallots

2 cloves garlic, minced

1 red chile pepper, seeded and minced

3 Tablespoons fresh Italian parsley, chopped

1 teaspoon balsamic vinegar

¼ teaspoon dried thyme

⅓ teaspoon dried oregano

½ cup extra virgin olive oil

Combine all the ingredients in a food processor, and blend.

Per Tbsp.: calories: 34, protein: 0 gm., carbohydrates: 0 gm., fat: 4 gm.

Mint Pesto

Makes 1½ cups

1 cup fresh mint leaves

¼ teaspoon dried marjoram

½ teaspoon fresh thyme leaves

¼ teaspoon shallots

¼ teaspoon fresh basil

1 garlic clove, finely minced

½ cup extra virgin olive oil

½ cup pine nuts

⅓ cup Parmesan cheese

Add the mint, marjoram, thyme, shallots, basil, and garlic to a food processor and grind. Add the olive oil, pine nuts, and Parmesan cheese, and mix. Serve.

Per Tbsp.: calories: 61, protein: 1 gm., carbohydrates: 0 gm., fat: 6 gm.

Pasta Sauce

Makes 4 cups

2 Tablespoons olive oil

1 large white onion, chopped

2½ cups boiling water

2 cups dried tomato halves

2 cloves garlic, sliced

1 teaspoon salt

¼ teaspoon freshly ground pepper

¼ cup fresh parsley leaves, chopped

1 teaspoon dried basil

1 teaspoon dried oregano leaves

1 Tablespoon lemon juice

Heat the olive oil in a large skillet over medium heat. Add the onion and sauté for 5 minutes. Combine the water and tomatoes in a medium bowl. Place ⅔ of the tomato mixture in a blender, add the garlic, and puree. Add the puree with the remaining tomatoes and other ingredients to skillet. Bring to a boil, reduce the heat, and simmer for 2 minutes. Serve hot over pasta.

Per ½ cup: calories: 73, protein: 1 gm., carbohydrates: 9 gm., fat: 3 gm.

Tofu Cream Spread

Makes about 1⅓ cups

½ pound tofu, dried

2 Tablespoons extra virgin olive oil

2 Tablespoons Dijon mustard

1 clove garlic, minced

1 Tablespoon lemon juice or balsamic vinegar

Combine the above ingredients in a blender until smooth.

Per 2 Tbsp.: calories: 52, protein: 2 gm., carbohydrates: 1 gm., fat: 5 gm.

Maher's Spaghetti Sauce

Makes about 3 cups

Serve with hot pasta.

1 Tablespoon olive oil

1 large white onion, chopped

1 large green pepper, chopped

4 cloves garlic, finely chopped

1 cup fresh mushrooms

6 black olives

1 cup water

1 (16 oz.) can tomato sauce

¼ teaspoon black pepper

¼ teaspoon rosemary

1 cup fresh parsley, finely chopped

In a large saucepan, heat olive oil on medium heat. Sauté the onion for 3 minutes, then add the green pepper and sauté for 5 more minutes. Add the garlic, mushrooms, and olives, and cook for 5 more minutes. Stir in the water, tomato sauce, black pepper, and rosemary, and cook for 10 minutes uncovered. Add the parsley and cook for 5 more minutes.

Per cup: calories: 139, protein: 3 gm., carbohydrates: 18 gm., fat: 6 gm.

Tomato and Basil Sauce

Makes about 2 cups

Serve with hot pasta.

1 Tablespoon olive oil

8 cloves garlic, finely chopped

2 Tablespoons dried oregano

¼ cup fresh basil, chopped

¼ teaspoon black pepper

1 (15 oz.) can tomato sauce

½ cup water

3 Tablespoons Parmesan cheese

Heat the olive oil in a small saucepan, and sauté the garlic on medium heat for 2 minutes. Add the oregano, basil, and black pepper, and sauté for 2 more minutes. Add the tomatoes and water, and simmer, uncovered, for 20 minutes. Ten minutes before the sauce is done, add the Parmesan cheese.

Per cup: calories: 184, protein: 5 gm., carbohydrates: 20 gm., fat: 9 gm.

Oriental Tofu Spread

Makes 2 ⅓ cups

This is a zesty spread for sandwiches, crackers, or as dip for fresh vegetables.

1 pound tofu, dried
3 Tablespoons lime juice
2 Tablespoons extra virgin olive oil
1 Tablespoon soy sauce
1 clove garlic, minced
1 slice fresh ginger, julienned

Puree all above ingredients in a blender until smooth, stopping the blender occasionally to combine the ingredients. Chill for 6 hours before serving.

Per Tbsp.: calories: 17, protein: 1 gm., carbohydrates: 0 gm., fat: 1 gm.

Tofu Vinaigrette

Makes approximately 2 cups

½ cup extra virgin olive oil
3 Tablespoons red wine vinegar
1 Tablespoon Balsamic vinegar
1 teaspoon Dijon mustard
½ cup soft tofu

Combine all the ingredients in a blender until smooth.

Per Tbsp.: calories: 33, protein: 0 gm., carbohydrates: 0 gm., fat: 3 gm.

Tofu Sauce

Makes about 1 cup

This makes an excellent dip for raw vegetables.

1 cup soft tofu, drained and dried

2 Tablespoons fresh parsley, chopped

2 Tablespoons fresh basil, chopped

½ teaspoon Chinese mustard powder

1 tablespoon olive oil

½ teaspoon dill weed

⅓ cup green onions, sliced

2 leaves fresh mint, chopped

2 cloves garlic, minced

Process the tofu in a blender until smooth. Stir in the rest of ingredients by hand. Cover and refrigerate for several hours; serve cold.

Per Tbsp.: calories: 20, protein: 1 gm., carbohydrates: 0 gm., fat: 1 gm.

Sandwiches

Baked Eggplant Sandwich

Serves 4

1 large eggplant

2 Tablespoons olive oil

8 oz. goat cheese

1 loaf whole wheat bread, unsliced, or 1 round focaccia

2 canned, roasted red peppers, rinsed, or 2 medium tomatoes, sliced

1 bunch watercress or basil

Preheat the oven to 500°, and oil a 9" x 13" baking pan. Slice the eggplant horizontally into ½" slices, brush the slices with the olive oil, and place in the pan; bake for 15 minutes. Slice the loaf of bread lengthwise, and spread with the goat cheese. Arrange the baked eggplant on the bottom half of the loaf. Top with the peppers or tomatoes, watercress, and the top of the bread loaf. Slice and serve.

Per serving: calories: 367, protein: 15 gm., carbohydrates: 30 gm., fat: 19 gm.

Tofu Sandwiches

Makes 2 sandwiches

- **½ pound firm tofu, crumbled**
- **1 Tablespoon extra virgin olive oil**
- **½ teaspoon curry powder**
- **1 small green onion, sliced**
- **2 small tomatoes, thinly sliced**
- **2 pita breads**

Combine all the ingredients except the breads in a small bowl. Open the pita pockets, fill with the tofu mixture, and serve.

Per sandwich: calories: 309, protein: 14 gm., carbohydrates: 33 gm, fat: 12 gm.

Pita Bread and Sprouts Sandwich

Makes 2 servings —A healthy lunch sandwich.

- **1 cup alfalfa or clover sprouts**
- **2 tomatoes, finely chopped**
- **1 Tablespoon fresh lemon juice**
- **1 Tablespoon extra virgin olive oil**
- **2 pita breads, cut in half**

Mix the sprouts, tomatoes, lemon juice, and olive oil. Fill the pita breads with this mixture, and serve.

Per serving: calories: 241, protein: 7 gm., carbohydrates: 34 gm., fat: 8 gm.

Falafel

Makes 12 half-sandwiches

1 cup falafel mix*
2 cups olive oil
6 pita breads, halved
1 cup Tahini Sauce, page 132
6 tomatoes, finely chopped
1 cup fresh green mint leaves, chopped

Prepare falafel mix as indicated on the package. Heat the olive oil in a medium sized pan, and fry falafel. Drain falafel and dry using clean napkins. Spread the tahini sauce inside the pita breads. Place falafel inside the pita breads, and add the tomatoes and mint. Serve right away.

** Available in all Middle-Eastern food stores and many natural foods stores.*

Per half-sandwich: calories: 236, protein: 10 gm., carbohydrates: 28 gm., fat: 7 gm.

Garlic Bread

Makes 4 servings

2 pita breads
5 cloves garlic, minced
½ teaspoon anise seeds
½ teaspoon dry oregano
2 Tablespoons olive oil

Preheat the oven to 300°. Slit the pita breads in half, making four circles. Mix the garlic with the anise, oregano, and olive oil, and spread this mixture on each pita circle. Place the breads on a cookie sheet, and bake for 10 minutes.

Per serving: calories: 140, protein: 3 gm., carbohydrates: 16 gm., fat: 7 gm.

Pan Bagnat

Serves 4

We first tasted this delicious sandwich in the old Italian-French town of Nice, France. It is ideal for picnics.

1 long, wide loaf of French bread

8 oz. goat cheese

3 Tablespoons Dijon mustard

1 roasted red pepper, sliced

1 red onion, thinly sliced

1 tomato, thinly sliced

½ cucumber, cut in long thin slices

12 black olives, chopped

½ green pepper, thinly sliced

3 Tablespoons extra virgin olive oil

3 cloves garlic, finely chopped

3 Tablespoons balsamic vinegar

Cut the loaf of bread in half, horizontally. Spread the goat cheese and mustard on both sides. Arrange all the vegetables on the bottom bread half. Mix the oil, garlic, and vinegar together, and drizzle over the vegetables. Cover with the top half of the loaf, wrap tightly in plastic, and refrigerate for 3 hours. Slice to serve.

Per serving: calories: 431, protein: 15 gm., carbohydrates: 33 gm., fat: 26 gm.

Sautéed Mushrooms on Whole Wheat Toast

Serves 4

4 garlic cloves, minced
3 Tablespoons olive oil
1 pound mushrooms caps
¼ cup shallots, chopped
½ cup dry white vermouth
½ cup fresh parsley, chopped
½ cup sun dried tomatoes
freshly ground black pepper, to taste
8 slices whole wheat bread

In a medium saucepan, warm the garlic in the oil for 1 minute. Chop the mushroom caps into 2" pieces. Add the mushrooms and shallots to the garlic, and cook for 5 minutes. Add the wine, parsley, tomatoes, and pepper, and cook over low heat for 10 minutes. Serve over whole wheat toast.

Per serving: calories: 515, protein: 17 gm., carbohydrates: 67 gm., fat: 12 gm.

APPENDIX

When cooking the recipes of this book, you will discover that herbs are a common ingredient in many of them. It should be of no surprise to you that most Mediterranean cuisines rely on herbs to add to their flavor and soul. Therefore, we felt it was crucial to include a short section on herbs. Dr. Ralph Spiegl, a native Californian and faithful herb grower and lover was kind enough to write this section for us.

GROWING YOUR OWN HERBS

Special Contribution by Ralph Spiegl, M.D.

Herbs are a diverse group of plants with ancient historical bonds with human usage. They have long been prized as medicinal aids, as fragrances (potpourri and sachets), and as seasonings for foods. Often the difference between an extraordinary dish and an easily forgotten dish is the seasoning. And often the seasoning is the imaginative use of herbs. Very few of us nowadays can go to the barnyard and gather fresh eggs, but almost all of us can step outside to an herb garden or a pot container and gather fresh herbs for a meal. Many interesting uses of herbs will be found in the following recipes.

Herbs are generally hardy and easy to grow. They can fit into almost any garden and are readily raised in planter boxes and small pots. Many tolerate a rather "lean" soil and even little water or care. They tend to be pest-resistant and heat and sun-loving, though a few enjoy part shade and a moist soil.

A small investment of time, money, and effort can supply a rich reward of fresh seasonings for a great variety of dishes: soups, salads, dressing, sauces, casseroles, and roasts.

The most commonly used herbs for the kitchen are rosemary, thyme, basil, oregano, parsley, and mints. Others include chives, dill, fennel, bay leaf, marjoram, savory, and other less common varieties.

There are a few generalizations about the raising of herbs (more complete information is readily available in gardening books, such as *Sunset Guide to Western Gardens*):

1. Fertilize herbs very little or not at all.
2. Water at soil level rather than from above.
3. Do not use sprays or insecticides until you have tried a mild soap solution spray—this is often adequate to eradicate less hardy pests.

The chart reproduced below acts as a very simple guide for some of the most common questions concerning the most frequently used kitchen herbs.

Before planting herbs, one must really consider goals and needs. Do you plan a formal herb garden? (There are many

Herb	Soil (+enriched / -lean)	Sun	Shade	Annual	Perennial	OK for Container
Basil	+	✓		✓	✓	✓
Thyme	-	✓			✓	✓
Rosemary	-	✓			✓	✓
Oregano	+/-	✓		✓		✓
Parsley	+/-		✓	✓		✓
Mint	+		✓		✓	✓

variations of this.) Do you wish just a few favorite herbs for kitchen use? Do you want color in foliage or flowering herbs? Where will they be in your garden? Plant selection, to some extent, depends upon location: for example, full sun, part shade, a mix with other plants, etc. What size plants can you use? Herbs vary in size from creepers a few inches tall to some hardy specimens five to six feet tall. Some are easily grown in pots and containers. Some are treated as annuals and some are perennials.

Once some basic decisions are made, you are ready to make a selection of plants. Here is a simple check list of considerations before you buy:

—**Location:** Sun or shade.

—**Garden:** Herbal garden or mixed with other plants, or container planting.

—**Use:** Cooking, fragrance, drying for potpourri or wreaths, etc.

When your basic planning is done, where to purchase? Most general nurseries carry a selection of herbs and purely herbal nurseries are available, if not locally, then by catalog for mail ordering. You will quickly learn that many varieties of almost every herb exist, and an experienced nurseryman will be of great help in making your choices. Weekend farmers' markets are often a wonderful source of plants and information.

Soil should, of course, be prepared before planting, and most all herbs do well in a moderately composted soil with good drainage. As a rule, herbs do not like their "feet wet" (exceptions such as mint and violets do exist). And as a further rule (and only that), over-watering is more harmful than missing watering a few days, or even a week of care. Once rooted in proper soil and with the right sun exposure, these useful plants are "survivors"! Perennials, such as thyme, tarragon, oregano, need to be cut back to eliminate dead wood in winter. New growth comes from the base of the plant.

Here are some general classifications that will help you decide which plants are appropriate for your needs:

Pot and container growing herbs —
Oregano, thyme, chives, basil, santolinia, parsley, rosemary.

Herbs for color contrast in garden—
Lavender (green-gray foliage and flowers white to purple), santolinia (silvery gray with yellow flowers), basil (many varieties, green to purple color), thyme (a variety of leaf size and color available), oregano.

Sun lovers—
Nasturtium, rosemary, sage, santolinia, thyme, oregano, calendula, dill, fennel.

Full to part sun—
Basil, mint, parsley, chives, lemon balm, bergamot.

Tall-growing herbs—
Dill and fennel (back of garden). Good against fences (to avoid wind damage). Lavender (fragrant, and dries well for potpourri), lamb's ears for bouquets.

Flowering herbs—
Nasturtium, calendula, scented geranium, bergmot.

Fragrant herbs—
Lemon verbena, lavender, fragrant geraniums, sage, bay, santolinia.

As in most any area of interest, start modestly. Learn from experience. Do not be intimidated by technology. These are very "forgiving" plants. Most herbs are best used in their fresh form, except for oregano which is superior when used dried. Nothing is more fragrant in the kitchen than newly picked basil or mint. But many herbs do dry well and can be kept for long periods in closed jars, retaining much of their flavor and fragrance. Learning the many subtle uses of herbs will be an on-going enrichment of your enjoyment of cooking.

AN HISTORIC LOOK AT OLIVE OIL

The olive tree, *Olea europea*, has been used by humans since prehistoric times, and fossils show evidence of its presence 20 million years ago in the lands surrounding the Mediterranean sea. Known to many of the ancients as the queen of all trees, olives were probably first cultivated in Palestine around the fourth millennium B.C. It is believed that during its Old Kingdom period (3000-2000 B.C.), Egypt imported olive oil from Palestine.

Before the Roman Empire, Greeks were the most dominant colonizers, and by 600 B.C. they brought olive oil—and probably the cultivation of the tree—to the coastal regions that are now parts of Italy, France, and Spain, and to the city-states of North Africa.

During the Golden Age of Greece, and ever since, there was a busy traffic in olive oil from one end of the Mediterranean to the other. Amphoras used as olive oil storage jars are one of the most common objects found in the remains of ancient Greek shipwrecks. Thus did olive oil become one of the primary foods of the area that has since become known as the cradle of civilization.

Olive trees are remarkable for their longevity. Trees as old as five hundred years are common, and it has been claimed that the trees of the garden of Gethsemane are the same which witnessed the agony of Christ. No matter how long its lifespan, it is certain that the olive tree contributed more to ancient cultures than any other tree or plant.

From the earliest times, olive oil has found many uses other than as food. The Egyptians used it for preservation of mummies, and throughout the ancient world medicinal uses were common—as therapy in the treatment of cancer, and for most common ailments as well, ranging from diarrhea to constipation, and from burns to baldness. Folk medicines sometimes have a basis in chemistry and physiology that has been verified in modern times. It is known today that olives contain sitosterol-d-glucoside, a compound with antitumor properties. Olive leaf extracts have antibacterial as well as antihypertensive properties, which may explain why ancient physicians found that its use resulted in general improvement in the health of their patients, long before blood pressure could be scientifically measured.

A wide variety of other uses was found for olive oil: for example, in religious ceremonies, as part of the act of consecration or anointment; as an ingredient in cosmetics, and as the fuel for lamps. Olive branches were used by Egyptians for decorative purposes and by Greeks as symbols of victory in the ancient Olympic Games. Later it came to symbolize peace as well as victory, a cultural practice carried over to modern times in the emblem of the United Nations.

The most important use of the products of olive trees was of course as food. Along with bread and wine, the oil was treasured by the ancients of the whole Mediterranean region, and that is still true today. Now this primary ingredient of the delicious cuisines of so many countries has emerged as by far the most healthful cooking oil one can possibly use.

BIBLIOGRAPHY

Beilin, Lawrence J. Diet and Life-style in Hypertension: Changing Perspectives. *Journal of Cardiovascular Pharmacology*. 16: S62-S66, 1990.

Bitterman, W.A., Farhadian, H., Abu Samra, C., Lerner, D., Amoun, H., Krapf, D., Makov, U., Environmental and Nutritional Factors Significantly Associated with Cancer of the Urinary Tract Among Different Ethnic Groups. *Urologic Oncology* 18(3):501-508, 1991.

Buiatti, E., Palli, D., Decarli, A., Amadori, D., Avellini, C., Bianchi, S., Bonaguri, C., Cipriani, F., Cocco, P., Giacosa, A., Marubini, E., Minacci, C., Puntoni, R., Russo, A., Vindigni, C., Fraumeni, J.F., Blot, W.J. A Case-Control Study of Gastric Cancer and Diet in Italy: II Association with Nutrients. *International Journal of Cancer* 45:896-901, 1990.

Burkitt, Denis. Are Our Commonest Diseases Preventable? *The Pharos*, 19-21, winter 1991.

Cholesterol and Your Heart. *American Heart Association*, 1989.

Cipriani, F., Buiatti, E., Palli, D. Gastric Cancer in Italy. *Italian Journal of Gastroenterology* 23:429-435, 1991.

Cohen , J.M. and M.J. *The Penguin Dictionary of Quotations*. Penguin, 1964.

Donaldson, A.N. The Relation of Protein Foods to Hypertension. *Calif West Med* 24: 328-330, 1926.

Duke, James A. *Medicinal Plants of the Bible*. Conch Magazine Ltd., New York, 1983.

Dwyer, J.T. Health Aspects of Vegetarian Diets. *American Journal of Clinical Nutrition* 48: 811-818, 1988.

Ferro-Luzzi, Anna, et al. Changing the Mediterranean Diet: Effects on Blood Lipids. *The Journal of Clinical Nutrition*, 40: 1027-1037, 1984.

Grundy, Scott M. Comparison of Monounsaturated Fatty Acids and Carbohydrates for Lowering Plasma Cholesterol. *New England Journal of Medicine*, 314:745-748, 1986.

Grundy, Scott M., and Andrea Bonanome. Workshop on Monounsaturated Fatty Acids. *Arteriosclerosis*, 7(6):644-648, November/December 1987.

Hamilton, Eva May Nunnelley, Eleanor Noss Whitney, Frances Sienkiewicz. *Nutrition: Concepts and Controversies*. West Publishing Company, 1991.

Hillhouse, A.L. *An Essay on the History and Cultivation of the European Olive-Tree*. Cellot, Paris, 1820.

Insel, Paul, and Walton T. Roth. *Core Concepts in Health*. Mayfield Publishing Company, California, 1991.

Keys, Ancel. Olive Oil and Coronary Heart Disease. *The Lancet*, 983-984, April, 1987.

Kirschenbauer, H.G. *Fats and Oils*. Reinhold Publishing Corporation, New York, 1944.

Klein, Maggie Blyth. *The Feast of the Olive*. Aris Books, California, 1983.

Long, Patricia. Olive Oil: Sweetheart of the Arteries. *Hippocrates*, 21-25, October 1991.

Martin-Moreno, J.M., Willett, W.C., Gorgojo, L., Banegas, J.R., Rodriguez-Artalejo, F., Fernandez-Rodriguez, J.C., Maisoneuve, P., Boyle, P. Dietary Fat, Olive Oil and Breast Cancer Risk. *International Journal of Cancer* 58:774-780, 1994.

Mattson, Fred H. A Changing Role for Dietary Monounsaturated Fatty Acids. *Journal of American Dietetic Association*, 89: 387-391, 1989.

Mensink, Ronald P., and Martijn B. Katan. Effect of Monounsaturated Fatty Acids Versus Complex Carbohydrates on High-Density Lipoproteins in Healthy Men and Women. *The Lancet*, 122-125, January 1987.

Mills, P.K., Beeson, L., Phillips, R.L., Fraser, G.E. Dietary Habits and Breast Cancer Incidence Among Seventh-Day Adventists. *Cancer* 64:582-590, 1989

Oliver, M F. Dietary Fat and Coronary Heart Disease. *British Heart Journal*, 58: 423-428, 1987.

Payne, Wayne A., and Dale B. Hahn. *Understanding Your Health*. Times Mirror/Mosby, College Publishing, 1988.

Public Health Reports, 105(4):377, July-August 1990.

Reddy, B.S., Dietary Fat and Colon Cancer: Animal Model Studies. *Lipids* 27:807-813, 1992.

Rose, D.P., Boyar, A.P., Wynder, E.L. International Comparisons of Mortality Rates for Cancer of the Breast, Ovary, Prostate, and Colon, and Per Capita Food Consumption. *Cancer* 58:2363-2371, 1986.

Sacks, Frank M. Blood Pressure in Vegetarians. *American Journal of Epidemiology*, 100:390-398, 1974.

Sacks, Frank M., and Walter W. Willett. More on Chewing the Fat: the Good Fat and the Good Cholesterol. *The New England Journal of Medicine*, 1740-1742, December 1991.

Sandy, Brent D. *The Production and Use of Vegetable Oils in Ptolemaic Egypt*. Scholar Press, Georgia, 1989.

Schlesinger, Sarah, and Barbara Earnest. *The Low Cholesterol Olive Oil Cookbook*. Villard Books, New York, 1990.

Silber, Earl N. Ischemic Heart Disease. *Heart Disease*. Macmillon Publishing Company, 1987.

Simic, B.S., et al. The Longitudinal Investigations of the Influence of Diets Different in Their Caloric Value, Quantity and the Content of Fat and Sodium Chloride on Blood Pressure, and the Incidence of Abnormal Electrocardiograms in Old People. *Acta Med Iugosl* 17: 154-174, 1963.

Strauss, Maurice B. *Familiar Medical Quotations*. Little, Brown and Company. Boston.

The Holy Bible. Oxford University Press, 1962.

Tilli, Maria Grazia. *Vino e Olio in Toscana*. Il Fiore, Firenze, 1985.

University of Haifa, Israel Oil Industry, and Dagon Museum. *Olive Oil in Antiquity: Israel and neighboring countries*. Conference 1987, Haifa, Israel.

US Department of Health and Human Services. *Health United States*, 1990.

Welsch, C.W., Dietary Fat, Calories, and Mammary Gland Tumorigenesis. *Exercise, Calories, Fat, and Cancer*. Plenum Press, New York, 1992.

Index

THE AUTHORS

Maher A. Abbas—

A native of Lebanon, Dr. Abbas grew up among hills covered with olive trees dating back to biblical times, overlooking the Mediterranean Sea. He pursued his undergraduate studies at Emory University (Atlanta, Georgia) where he obtained two degrees in biology and chemistry, and graduated with several academic honors, including Phi Beta Kappa. Dr. Abbas was awarded his medical degree from Stanford University Medical School in Palo Alto, California. He spent two and a half years researching cardiovascular disease at the Falk Cardiovascular Research Center of Stanford University and was awarded the American Federation Clinical Research Award for his contribution to the understanding of cardiovascular disease. Along with Mrs. Spiegl, he has organized and taught several classes, seminars, and workshops at Stanford University and in the Bay area on healthful diet and hearty cooking. Among his many accomplishments is his participation in the 400 m and 800 m running events in the 1988 Olympic Games in Seoul, Korea.

Marilyn J. Spiegl —

Mrs. Spiegl was born in Chicago, Illinois. She attended Northwestern University (Evanston, Illinois) as an education major and studied design in San Francisco at the Rudolph Schaefer School of Design. As an interior designer and art advisor, Mrs. Spiegl has been consultant on prestigious projects to architectural firms as well as corporate designer of a major California corporation. Cooking and painting are her avid interests and Mrs. Spiegl has studied in France, Italy, and the United States with notable chefs. She is the mother of three fine young adults, and is most happily married to Dr. Ralph Spiegl—hence her considerable involvement in promoting healthful life-styles.

Healthy World Cuisine!

Delicious Jamaica $11.95

Flavors of India $12.95

A Taste of Mexico $13.95

From a Traditional Greek Kitchen $12.95

Good Time Eatin' in Cajun Country $9.95

Indian Vegetarian Cooking $12.95

Flavors of the Southwest $12.95

COMING SOON!
Flavors of Korea $12.95
and
Nonna's Italian Kitchen $14.95

From the Global Kitchen $11.95

Ask your store to carry these books, or you may order directly from:

The Book Publishing Company P.O. Box 99, Summertown, TN 38483
Or call: 1-800-695-2241. Please add $2.50 per book for shipping.